T0249493

Three-Dimensional Integrated Circuit Design

Three-Dimensional Integrated Circuit Design

Vasilis F. Pavlidis

Eby G. Friedman

AMSTERDAM • BOSTON • HEIDELBERG • LONDON
NEW YORK • OXFORD • PARIS • SAN DIEGO
SAN FRANCISCO • SINGAPORE • SYDNEY • TOKYO

ELSEVIER

Morgan Kaufmann Publishers is an imprint of Elsevier

MORGAN KAUFMANN

Morgan Kaufmann Publishers is an imprint of Elsevier.
30 Corporate Drive, Suite 400, Burlington, MA 01803, USA

This book is printed on acid-free paper.

Library of Congress Cataloging-in-Publication Data
Application Submitted

ISBN: 978-0-12-374343-5

For information on all Morgan Kaufmann publications,
visit our Web site at www.mkp.com or www.elsevierdirect.com

Printed and bound in the United Kingdom
Transferred to Digital Printing, 2011

The Morgan Kaufmann Series in Systems on Silicon

Series Editor
Wayne Wolf
Georgia Institute of Technology

To the memory of my beloved grandfathers, Athanasios Theophilidis and Vasileios Pavlidis

To my wife and lifelong companion, Laurie. LL

Contents

Preface

The seminal reason for this book is the lack of a unified treatment of the design of three-dimensional integrated circuits despite the significant progress that has recently been achieved in this exciting new technology. Consequently, the intention of this material is to cohesively integrate and present research milestones from different, yet interdependent, aspects of three-dimensional integrated circuit design. The foremost goal of the book is to propose design methodologies for 3-D circuits; methodologies that will effectively exploit the flourishing manufacturing diversity existing in three-dimensional integration. While the focal point is design techniques and methodologies, the material also highlights significant manufacturing strides that complete the research mosaic of three-dimensional integration.

Three-dimensional or vertical integration is an exciting path to boost the performance and extend the capabilities of modern integrated circuits. These capabilities are inherent to three-dimensional integrated circuits. The former enhancement is due to the considerably shorter interconnect length in the vertical direction and the latter is due to the ability to combine dissimilar technologies within a multi-plane system. It is also worth noting that vertical integration is particularly compatible with the integrated circuit design process that has been developed over the past several decades. These distinctive characteristics make three-dimensional integration highly attractive as compared to other radical technological solutions that have been proposed to resolve the increasingly difficult issue of on-chip interconnect.

The opportunities offered by three-dimensional integration are essentially limitless. The constraints stem from the lack of design and manufacturing expertise for these circuits. The realization of these complex systems requires advanced manufacturing methods and novel design technologies across several abstraction levels. The development of these capabilities will be achieved if the physical behavior and mechanisms that govern interplane communication and manufacturing are properly understood. The focus of this book diligently serves this purpose.

This book is based upon the body of research carried out by Vasilis Pavlidis from 2002 to 2008 at the University of Rochester during his doctoral study under the supervision of Professor Eby G. Friedman. Recognizing the importance of the vertical interconnections in 3-D circuits, these structures are central to the content of this book. Tutorial chapters are dedicated to manufacturing processes, technological challenges, and electrical models of these structures. The vertical wires are investigated not only as a communication medium but also as heat conduits within the 3-D system. From this perspective, novel and efficient algorithms are presented for improving the signal propagation delay of heterogeneous three-dimensional systems that consider the electrical behavior of the vertical interconnects. Additionally, the important role that the vertical interconnects play in global signaling and thermal amelioration is described. Measurements from a case study of a 3-D circuit increase the physical understanding and intuition of this critical interconnect structure.

The short vertical wires enable several 3-D architectures for communication centric circuits. Opportunities for improved communication bandwidth, enhanced latency, and low power in 3-D systems are investigated. Analytic models and exploratory tools for these architectures are also discussed. The intention of this part of the book is to illuminate important design issues while providing guidelines for designing these evolving 3-D architectures.

The organization of the book is based on a bottom-up approach with technology and manufacturing of 3-D systems as the starting point. The first two chapters cover the wide spectrum of available 3-D technologies and related fabrication processes. These chapters demonstrate the multi-technology palette that is available in designing a 3-D circuit. Based on these technologies and *a priori* interconnect models, projections of the capabilities of 3-D circuits are provided. Physical design methodologies for 3-D circuits, which are the core of this book, are considered in the next eight chapters. The added design complexity due to the multi-plane nature of a 3-D system, such as placement and routing, is reviewed. Efficient approaches to manage this complexity are presented. Extensions to multi-objective methodologies are discussed with the thermal issue as a primary reference point. As the vertical interconnects can behave either obstructively or constructively within a 3-D circuit, different design methodologies to utilize these interconnects are extensively discussed. Specific emphasis is placed on the through silicon vias, the important vertical interconnection structure for 3-D circuits. Various 3-D circuit architectures, such as a processor and memory system, FPGAs, and on-chip networks, are discussed. Different 3-D topologies for on-chip networks and FPGAs are explored. Novel algorithms and accurate delay and power models are reviewed. The important topic of synchronization is also investigated, targeting the challenges of distributing a clock signal throughout a multi-plane circuit. Experimental results from a 3-D fabricated circuit provide intuition into this global signaling issue.

Three-dimensional integration is a seminal technology that will prolong the semiconductor roadmap for several generations. The third dimension offers nonpareil opportunities for enhanced performance and functionality; vital requirements for contemporary and future integrated systems. Considerable progress in manufacturing 3-D circuits has been achieved within the last decade. 3-D circuit design methodologies, however, considerably lag these technological advancements. This book ambitiously targets to fill this gap and strengthen the design capabilities for 3-D circuits without overlooking aspects of the fabrication process of this emerging semiconductor paradigm.

ACKNOWLEDGMENTS

The authors are thankful to Professor Dimitrios Soudris of Democritus University of Thrace for his important contributions to Chapter 9. The authors are also grateful to Dr. Sankar Basu of the National Science Foundation. Without his support, much of this research would not have been possible.

The authors would also like to thank the MIT Lincoln Laboratories for providing foundry support for the 3-D test circuit and, more specifically, to Dr. Chenson Chen and Bruce Wheeler of MIT Lincoln Laboratories for their assistance during the 2nd multi-project run. We would also like to thank Yunliang Zhu, Lin Zhang, and Professor Hui Wu of the University of Rochester for their help during the testing of the 3-D circuit. Special thanks to Nopi Pavlidou for the creative design of the front cover of the book. Finally, this book would not have been possible without the continuous support and encouragement of Charles Glaser, who welcomed the idea of this book and prodded us throughout the writing process. Thanks, Chuck!

The original research work presented in this book was made possible in part by the National Science Foundation under Contract No. CCF-0541206, grants from the New York State Office of Science, Technology & Academic Research to the Center for Advanced Technology in Electronic Imaging Systems, and by grants from Intel Corporation, Eastman Kodak Company, and Freescale Semiconductor Corporation.

Vasilis F. Pavlidis and Eby G. Friedman
Rochester, New York

ACKNOWLEDGMENTS

Introduction

Electronics has experienced tremendous growth over the past century. During this time period, the use of electronic merchandise has steadily increased while the size of these products has decreased. The transition from micro- to nano-scale has further enhanced the applicability of electronic products into new areas, which were previously precluded due to the prohibitive size of these systems. The seed that caused this growth — the "big-bang" of electronics — was a grain of germanium on which the point contact transistor was fabricated for the first time in 1947 by J. Bardeen, W. Brattain, and W. Shockley [1].

This invention appealed to the interests of scientists and engineers, resulting in research efforts focused on developing semiconductor devices to replace the bulky, power-hungry, and low-performance vacuum tubes and electromechanical relays on which much of the electronics of the pre-transistor era were based [2]. This quest led to the emergence of a new branch of electronics, namely, the semiconductor industry, which experienced tremendous growth over the following decades. An important engine behind this explosive evolution was the plethora of semiconductor-based applications. Semiconductor products have ultimately affected every component of our society.

For example, manufacturing was considerably advanced due to automation, reducing the cost and time to market and offering more reliable products, while increasing employee safety through highly sophisticated electronic security systems that can recognize and warn of critical equipment failures [3]. Office automation also significantly simplified the painstaking process of writing letters, memos, and reports, while facilitating filing by offering a variety of storage media [4]. In addition, medical procedures were simplified, and invasive diagnostic methods, which were often dangerous to patients, were replaced by safe and effective noninvasive techniques. Furthermore, minute electronics devices, such as the pacemaker and hearing aid, enhanced the treatment of myriad patients [5]. Communications is another component of our society that has passed through a revolution during the past several decades. Satellite communications, geographical positioning systems, cell phones, and the Internet are some of the salient achievements in the communications field. Without the robust and powerful transistor, most of these capabilities would be rather primitive, if not impossible.

Science and engineering perhaps benefited the most from the micro-electronics revolution and the flourishing semiconductor industry [6]. Computational tasks, once formidable, are now solved in fractions of a second. Numerous computer programs, measurement instrumentation, and observation apparatus have been developed, expanding our knowledge of the environment, nature, and the universe. Information processing and propagation capabilities have been improved by orders of magnitude as compared to the beginning of the twentieth century, making knowledge available in almost any place around the globe. The electronics industry has been, in turn, assisted by these developments, both intellectually and financially. Novel applications further boosted the semiconductor revolution, producing colossal revenues that helped establish and fuel industrial R&D. Some of the milestones of this stupendous progress are described in the following section. Interconnect-related problems since the earliest days of the integrated circuit industry and the impending performance bottleneck caused by the interconnect are discussed in Section 1.2. A promising solution and an important next step in the evolution of the microelectronics field, namely, three-dimensional integration, is introduced in Section 1.3. An outline of this book is presented in Section 1.4.

1.1 FROM THE INTEGRATED CIRCUIT TO THE COMPUTER

During the years that followed the genesis of the point contact transistor in 1947, several different types of semiconductor devices were fabricated to satisfy a variety of important applications in control systems, military, medicine, and a host of other areas. A time line of these inventions is shown in Figure 1-1. These transistors were discrete components connected together with traces of metal that implemented different circuit functions. Although these innovative devices could perform better and more reliably at greater frequencies than vacuum tubes or other electromechanical equipment, a system that exclusively consisted of discrete components would be of limited performance and could not exploit the

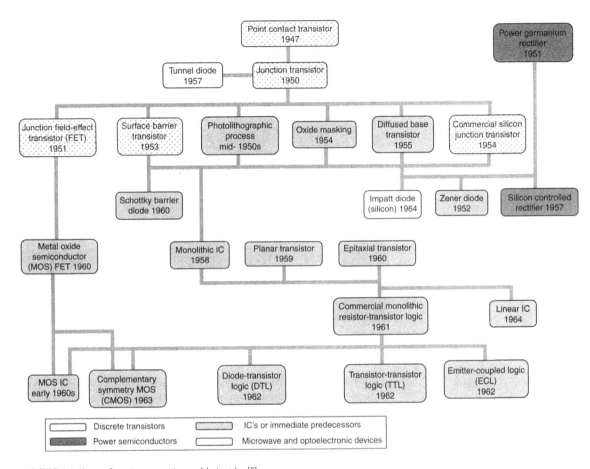

■ FIGURE 1-1 History of transistor generations and logic styles [8].

full potential of semiconductor devices. Not until the development of the planar process in 1960, which resulted in the familiar integrated circuit, as shown in Figure 1-2, could the capabilities of the transistor begin to be fully utilized [7]. This achievement reinforced the growth of the semiconductor industry, offering a large number of devices integrated within the silicon, implementing a vast variety of circuit functions.

Later improvements in the fabrication process and integration on inexpensive silicon resulted in integrated circuits (ICs) with higher yield and hence lower cost, greater performance, and enhanced reliability. Within the next decade, several logic-style families such as transistor-transistor logic (TTL), emitter-coupled logic (ECL), and complementary metal oxide semiconductor (CMOS) were proposed. As the complexity

■**FIGURE 1-2** The first planar integrated circuit [9].

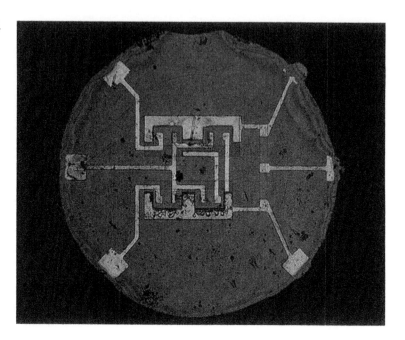

of the ICs increased, a profound need developed for circuits suitable for generalized applications. Indeed, thus far, each IC was designed to serve a single application, requiring companies to design a variety of low-cost components to maintain profitability. The answer to this profound need was provided by M. E. Hoff Jr., an engineer at Intel. He envisioned a more flexible way to utilize the capabilities of these integrated circuits. Encouraged by the founders of Intel, Gordon Moore and Robert Noyce, the result of this effort was the first microprocessor, namely, the 4004, which is illustrated in Figure 1-3. This 0.11×0.15 square inch IC could execute addition and multiplication with four-bit numbers, while a bank of registers was used for storage purposes. Although this capability seems trivial today, the 4004 microprocessor fundamentally altered the way that computers were perceived and used.

Since the 4004 was announced in 1971, microprocessors and ICs in general have steadily improved, demonstrating higher performance and reliability. This fascinating trajectory was essentially driven by the maturation of the semiconductor manufacturing process, supported by the continued scaling of the transistors. The merits of this evolution were foreseen quite early by Moore and Noyce, as discussed in the following section, where the physical limitations of technology scaling are also described.

1.2 **INTERCONNECTS, AN OLD FRIEND**

During the infancy of the semiconductor industry, the connections among the active devices of a circuit presented an important obstacle for increasing circuit performance. The significant capacitance of the interconnects necessitated large power drivers and hindered a rapid increase in performance that could be achieved by the transistors. Noyce had already noticed the importance of the interconnects, such as the increase in delay and noise due to coupling with neighboring interconnects [10]. The invention of the integrated circuit considerably alleviated these early interconnect-related problems by bringing the interconnects on-chip. The interconnect length was significantly reduced, decreasing the delay and power consumption while reducing the overall cost. From a performance point of view, the speed of the transistors dominated the overall delay characteristics. Over the next three decades, on-chip interconnects were not the major focus of the IC design process, as performance improvements reaped from scaling the devices were much greater than any degradation caused by the interconnects.

With continuous technology scaling, however, the interconnect delay, noise, and power grew in importance [11], [12]. A variety of methodologies at the architectural, circuit, and material levels has been developed to address these interconnect design objectives. At the material level, manufacturing innovations such as the introduction of copper interconnects and low-k dielectric materials helped to prolong the improvements in performance gained from scaling [13]–[17]. This situation is due to the lower resistivity of the copper as compared to aluminum interconnects and the lower dielectric permitivity of the new insulator materials as compared to silicon dioxide.

Multitier interconnect architectures [18], [19], shielding [20], wire sizing [21], [22], and repeater insertion [23] are only a handful of the many methods employed to cope with interconnect issues at the circuit level. Multitier interconnect architectures, for example, support tiers of metal layers with different cross sections [19], as illustrated in Figure 1-4. Each tier typically consists of multiple metal layers routed in orthogonal directions with the same cross section. The key idea of this structure is to utilize wires of decreasing resistance to connect those circuits located farther away. Thus, the farther the distance among the circuits, the thicker the wires used to connect these circuits. The increase in the cross section of the wires is shown in Figure 1-4. The thickness of the tiers, however, is limited by the fabrication technology and related reliability and yield concerns.

Varying the width of the wires, also known as wiring sizing, is another means to manage the interconnect characteristics. Wider wires lower the interconnect resistance, decreasing the attenuative properties of the wire. Although wire sizing typically has an adverse effect on the power consumed by the interconnect, proper sizing techniques can also decrease the power consumption [22], [24].

Other practices do not modify the physical characteristics of the propagation medium. Rather, by introducing additional circuitry and wire resources, the performance and noise tolerance of an interconnect system can be enhanced. For instance, in a manner similar to the use of repeaters in telephone lines systems, a properly designed interconnect system with buffers (also known as repeaters) amplifies the attenuated signals, recovering the originally transmitted signal that is propagated along a line. Repeater insertion effectively converts the square dependence of the delay on the interconnect length to a linear function of length, as shown in Figure 1-5.

Shielding is an effective technique to reduce crosstalk among adjacent interconnects. Single- or double-sided shielding, as depicted in Figure 1-6, is commonly utilized to improve signal integrity. Shields can also improve interconnect delay and power, particularly in buss

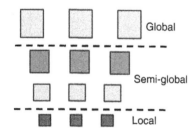

FIGURE 1-4 Interconnect architecture including local, semi-global, and global tiers. The metal layers on each tier are of different thicknesses.

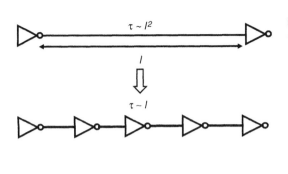

■ **FIGURE** 1-5 Repeaters are inserted at specific distances to improve the interconnect delay.

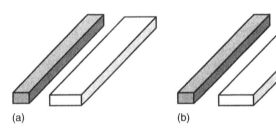

(a)   (b)

■ **FIGURE** 1-6 Interconnect shielding to improve signal integrity: (a) single-sided and (b) double-sided shielding. The shield and signal lines are illustrated by the grey and white color, respectively.

architectures, in addition to mitigating noise. Careful tuning of the relative delay of the propagated signals [25] and signal encoding schemes [26] are other strategies to maintain signal integrity. Despite the benefits of these techniques, issues arise such as the increase in power consumption, greater routing congestion, the reduction of wiring resources, and an increase in area.

At higher abstraction levels, pipelining the global interconnects and employing error-correction mechanisms can partially improve the performance and fault tolerance of the wires. The related cost of these architecture level techniques in terms of area and design complexity, however, increases considerably. Other interconnect schemes such as current mode signaling [27], wave pipelined interconnects [28], and low swing signaling [29] have been proposed as possible solutions to the impending interconnect bottleneck. These incremental methods, however, have limited ability to reduce the length of the wire, which is the primary cause of the deleterious behavior of the interconnect.

Novel design paradigms are therefore required that do not impede the well-established and historic improvement in performance in next-generation integrated circuits. Canonical interconnect structures that utilize Internet-like packet switching for data transfer [30], optical interconnects [31], and three-dimensional integration are possible solutions for providing communication among devices or functional blocks within an IC.

On-chip networks can considerably enhance the communication bandwidth among the individual functional blocks of an integrated system, since each of these blocks utilizes the resources of the network. In addition, noise issues are easier to manage as the layered structure of the communication protocols utilized within on-chip networks provides error correction. The speed and power consumed by these networks, however, are eventually limited by the delay of the wires connecting the network links.

Alternatively, on-chip optical interconnects can greatly improve the speed and power characteristics of interconnects within an integrated circuit, replacing the critical electrical nets with optical links [32], [33]. On-chip optical interconnects, however, remain a technologically challenging problem. Indeed, integrating a modulator and detector onto the silicon within a standard CMOS process is a difficult task [31]. In addition, the detector and modulator should exhibit performance characteristics that ensure the optical links outperform the electrical interconnects [33]. Furthermore, an on-chip optical link consumes a larger area as compared to a single electrical interconnect line. To limit the area consumed by the optical interconnect, multiplexing the optical signals (wavelength division multiplexing (WDM)) can be exploited. On-chip WDM, however, imposes significant challenges.

Volumetric integration by exploiting the third dimension greatly improves the interconnect performance characteristics of modern integrated circuits, while the interconnect bandwidth is not degraded. In general, three-dimensional integration should not be seen as competitive but rather synergistic with on-chip networks and optical interconnections. The unique opportunities that three-dimensional integration offers to the circuit design process and the challenges that arise from the increasing complexity of these systems are discussed in the following section.

1.3 THREE-DIMENSIONAL OR VERTICAL INTEGRATION

Successful fabrication of vertically integrated devices dates back to the early 1980s [34]. The structures include 3-D CMOS inverters where the positive-channel metal oxide semiconductor (PMOS) and negative-channel metal oxide semiconductor (NMOS) transistors

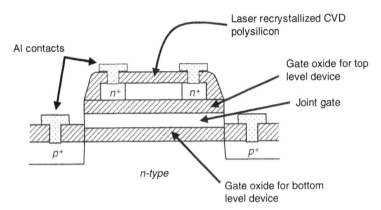

Al contacts

Laser recrystallized CVD polysilicon

Gate oxide for top level device

Joint gate

n^+ n^+

p^+ p^+

n-type

Gate oxide for bottom level device

■ **FIGURE 1-7** Cross section of a JMOS inverter [35].

share the same gate, considerably reducing the total area of the inverter, as illustrated in Figure 1-7. The term joint metal oxide semiconductor (JMOS) was used for these structures to describe the joint use of a single gate for both devices [35]. In the following years, research on three-dimensional integration remained an area of limited scientific interest. Due to the increasing importance of the interconnect and the demand for greater functionality on a single substrate, vertical integration has recently become a more prominent research topic. Over the last five years, three-dimensional integration has evolved into a design paradigm manifested at several abstraction levels, such as the package, die, and wafer levels. Alternatively, different manufacturing processes and interconnect schemes have been proposed for each of these abstraction levels [36]. The salient features and important challenges for three-dimensional systems are briefly reviewed in the following subsections.

1.3.1 Opportunities for Three-Dimensional Integration

The quintessence of three-dimensional integration is the drastic decrease in the length of the longest interconnects across an integrated circuit. To illustrate this situation, consider the simple example structure shown in Figure 1-8. A common metric to characterize the longest interconnect is to assume that the length of a long interconnect is equal to twice the length of the die edge. Consequently, assuming a planar integrated circuit with an area A, the longest interconnect in a planar IC has a length $L_{\mathrm{max},2-D} = 2\sqrt{A}$. Implementing the same circuitry onto two bonded dies requires an area $A/2$ for each plane, while the total area of the system remains the same.

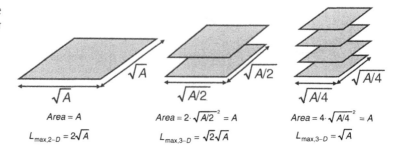

$Area = A$

$L_{max,2-D} = 2\sqrt{A}$

$Area = 2 \cdot \sqrt{A/2}^2 = A$

$L_{max,3-D} = \sqrt{2}\sqrt{A}$

$Area = 4 \cdot \sqrt{A/4}^2 = A$

$L_{max,3-D} = \sqrt{A}$

Hence, the length of the longest interconnect for a two-plane 3-D IC is $L_{max,3-D} = 2\sqrt{A/2}$. Increasing the number of dies within a 3-D IC to four, the area of each die can be further reduced to $A/4$, and the longest interconnect would have a length of $L_{max,3-D} = 2\sqrt{A/4}$. Consequently, the wire length exhibits a reduction proportional to $\sqrt{n}$, where n is the number of dies or physical planes that can be integrated into a 3-D system. Although in this simplistic example the effect of the connections among circuits located on different dies is not considered, *a priori* accurate interconnect prediction models adapted for 3-D ICs also demonstrate a similar trend caused by the reduction in wirelength [37]. This considerable decrease in length is a promising solution for increasing the speed while reducing the power dissipated by an IC.

Another characteristic of 3-D ICs of greater importance than the decrease in the interconnect length is the ability of these systems to include disparate technologies. This defining feature of 3-D ICs offers unique opportunities for highly heterogeneous and multifunctional systems. A real-time image processing system in which the image sensor on the topmost plane captures the light, the analog circuitry on the plane below converts the signal to digital data, and the remaining two planes of digital logic process the information from the upper planes is a powerful example of a heterogeneous 3-D system-on-chip (SoC), with considerably improved performance as compared to a planar implementation of the same system [38], [39]. Another example where the topmost plane can include other types of sensors such as seismic and acoustic sensors and an additional plane with wireless communications circuitry is schematically illustrated in Figure 1-9. A vast pool of applications, in the military, medical, and wireless communication domains, as well as low-cost consumer products exists for vertical integration, as the proximity of the system components due to the third dimension is suitable for either the high-performance or low-power ends of the SoC application spectrum.

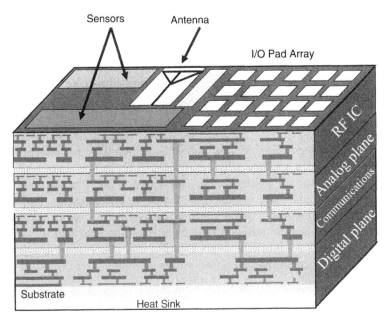

Sensors Antenna

I/O Pad Array

Substrate

Heat Sink

■ **FIGURE 1-9** An example of a heterogeneous 3-D system-on-chip comprising sensor and processing planes.

1.3.2 **Challenges for Three-Dimensional Integration**

Developing a design flow for 3-D ICs is a complicated task with many ramifications. A number of challenges at each step of the design process have to be satisfied for 3-D ICs to successfully evolve into a mainstream technology. Design methodologies at the front end and mature manufacturing processes at the back end are required to effectively provide large-scale 3-D systems. Several of the primary challenges to the successful development of 3-D systems are summarized below.

Technological/Manufacturing Limitations

Some of the fabrication issues encountered in the development of 3-D systems concern the reliable lamination of several ICs, possibly from dissimilar technologies. The stacking process should not degrade the performance of the individual planes, while guaranteeing the planes remain bonded throughout the lifetime of the 3-D system. Furthermore, packaging solutions that accommodate these complicated 3-D structures need to be developed. In addition, the expected reduction in wirelength depends on the vertical interconnects utilized to propagate signals and power throughout the planes of a 3-D system.

The technology of the interplane interconnects is a primary determining factor in circuit performance. Consequently, providing high-quality and highly dense vertical interconnects is of fundamental importance in 3-D circuits; otherwise, the performance or power gains achieved by the third dimension will diminish [40]. Alternatively, the density of this type of interconnect dictates the granularity at which the planes of the system can be interconnected, directly affecting the bandwidth of the interplane communication.

Testing

Manufacturing a 3-D system typically includes bonding multiple physical planes. The stacking process can occur either in a wafer-to-wafer, die-to-wafer, or die-to-die manner. Consequently, novel testing methodologies at the wafer and die level are required. Developing testing methodologies for wafer-level integration is significantly more complicated than die-level testing techniques. The considerable reduction in turnaround time due to the higher integration, however, may justify the additional complexity of these testing methods.

An important distinction between 2- and 3-D IC testing is that in the latter case only part of the functionality of the system is tested at a given time (typically, only one plane is tested at a time). This characteristic requires additional resources, for example, scan registers embedded within each plane. Furthermore, additional interconnect resources, such as power/ground pads, are necessary. These extra pads supply power to the plane during testing. In general, testing strategies for 3-D systems should not only include methodologies for generating appropriate input patterns for each plane of the system, but also minimize the circuitry dedicated to efficiently test each plane within a 3-D stack.

Interconnect Design

Design for test strategies are only a portion of the many design methodologies that need to be developed for 3-D ICs. Interconnect design and analysis of 3-D circuits are also challenging tasks. This capability is primarily due to the inherent heterogeneity of these systems, where different fabrication processes or disparate technologies are combined into a 3-D circuit. Consequently, models that consider the particular traits of 3-D ICs are necessary. In these diverse systems, global interconnect, such as clock and power distribution, grow in importance. Furthermore, well-developed noise mitigation techniques may not be suitable for 3-D circuits. Noise caused by capacitive and inductive coupling of the interconnections between adjacent planes needs to be considered from a 3-D perspective [41]. For example, signal switching on the topmost metal layer of a digital plane can produce a noise spike in an adjacent analog plane bonded in a front-to-front fashion with the digital plane. Considering the different forms of 3-D integration and the various manufacturing approaches, the development of interconnect design techniques and methodologies is a primary focus in high-performance 3-D systems.

Thermal Issues

A fundamental concern in the design of 3-D circuits is thermal effects. Although the power consumption of these circuits is expected to decrease due to the considerably shorter interconnects, the power density will greatly increase since there is a larger number of devices per unit area as compared to a planar 2-D circuit. As the power density increases, the temperature of those planes not adjacent to the heat sink of the package will rise, resulting in degraded performance, or accelerated wear-out mechanisms. Exploiting the performance benefits of vertical integration while mitigating thermal effects is a difficult task. In addition to design practices, packaging solutions and more effective heat sinks are additional approaches to alleviate thermal effects.

CAD Algorithms and Tools

Other classic problems in the IC design process, such as partitioning, floorplanning, placement, and routing, will need to be revisited in an effort to develop efficient solutions that can support the complexity of three-dimensional systems. To facilitate the front-end design process, a capability for exploratory design is also required. For example, design entry tools that provide a variety of visualization options can assist the designer in more easily comprehending and managing the added complexity of these circuits. In addition, as diverse technologies are combined within a single 3-D system, algorithms that include behavioral models for a larger variety of components are needed. Furthermore, the computational power of the simulation tools will need to be significantly extended to ensure that the entire system can be efficiently evaluated.

In this book, emerging 3-D technologies and design methodologies are analyzed, and solutions for certain critical problems are discussed. The next section presents an outline of the book.

1.4 **BOOK ORGANIZATION**

A brief description of the challenges that 3-D integration faces is provided in the previous section. Several important problems are considered and innovative solutions are presented throughout this book. In the next chapter, various forms of vertically integrated systems are discussed. Different implementations of three-dimensional circuits at the package and die integration levels are reviewed. Some of these approaches, such as a wire bonded stacked die, have become commercially available, while others have not yet left the R&D phase. Although vertical integration of packaged or bare die offers substantial improvements over planar multichip packaging solutions, the increasing number of inputs and outputs (I/Os) hampers potential performance advancements. This situation is primarily due to manufacturing limitations hindering aggressive scaling of the off-chip interconnects in order to satisfy future I/O requirements.

Consequently, in the third chapter, emphasis is placed on those technologies that enable 3-D integration, where the interconnections between the non-coplanar circuits are achieved through the silicon by short vertical vias. These interconnect schemes provide the greatest reduction in wire length and, therefore, the largest improvement in performance and power consumption. Specific fabrication processes that have been successfully developed for three-dimensional circuits are reviewed.

A theoretical analysis of interconnections in 3-D ICs is offered in the fourth chapter. This investigation is based on *a priori* interconnect prediction models. These typically stochastic models are used to estimate the length of the on-chip interconnects. The remaining sections of this chapter apply the interconnect distributions to demonstrate the opportunities and performance benefits of vertical integration.

The following four chapters focus on issues related to the physical design of 3-D ICs. The complexity of the 3-D physical design process is discussed in Chapter 5. Several approaches for classical physical design issues, such as floorplanning, placement, and routing, from a 3-D perspective, are reviewed. Other important issues, such as the allocation of decoupling capacitance, are also presented.

In the sixth chapter, physical design techniques for 3-D ICs are extended to thermal design and management. Modeling and design methodologies of thermal effects are discussed. Design techniques that utilize additional interconnect resources to increase the thermal conductivity within a multiplane system are emphasized.

Beyond the reduction in wirelength that stems from three-dimensional integration, the delay of those interconnects connecting circuits located on different physical planes of a 3-D system (*i.e.*, the interplane interconnects) can be further improved by optimally placing the through silicon vertical vias. Considering the highly heterogeneous nature of 3-D ICs that results from the nonuniform impedance characteristics of the interconnect structures, a methodology is described in Chapter 7 to minimize the delay of the interplane interconnects. An interconnect line that includes only one through silicon via is initially investigated. The location of the through silicon via that minimizes the delay of a line is analytically determined. The degradation in delay due to the nonoptimum placement of the 3-D vias is also discussed. In order to incorporate the presence of physical obstacles, such as logic cells and pre-routed interconnects (for example, segments of the power and clock distribution networks), the discussion in this chapter proceeds with interconnects that include more than one through silicon via. An accurate heuristic is described to implement an efficient algorithm for placing the through silicon vias to minimize the overall delay of a multiplane interconnect.

By extending the heuristic for two-terminal interconnects, a near-optimal heuristic for multiterminal nets in 3-D ICs is described in Chapter 8. Necessary conditions for locating the through silicon vias are described. An algorithm that exhibits low computational time is also presented. The improvement in delay that can be achieved by placing the through silicon vias for different via placement scenarios is investigated. For the special case where the delay of only one branch of a multiterminal net is minimized, a simpler optimization procedure is described. Based on this approach, a second algorithm is presented. Finally, the sensitivity of this methodology to the interconnect impedance characteristics is demonstrated, depicting a significant dependence of the delay on the interconnect parameters.

Exploiting the advantages of 3-D integration requires the development of novel circuit architectures. The 3-D implementation of a microprocessor-memory system is a characteristic example of the architectures discussed in Chapter 9. Major improvements in throughput, power consumption, and cache miss rate are demonstrated. Communication centric architectures, such as networks-on-chip (NoC), are also discussed. On-chip networks are an important design paradigm to appease the interconnect bottleneck, where information is communicated among circuits within packets in an Internet-like fashion. The synergy between these two design paradigms, NoC and 3-D ICs, can be exploited to significantly improve performance while decreasing the power consumed in future communication limited systems.

As noted earlier, the distribution of the clock signal in a 3-D IC is an important and difficult task. In Chapter 10, a variety of clock networks, such as H-trees, rings, tree-like networks, and trunk based networks, are explored in terms of clock skew and power consumption to determine an effective clock distribution network for 3-D ICs. A prototype test circuit composed of these networks has been designed and manufactured with the 3-D fabrication process developed at MIT Lincoln Laboratories. A description of the design process and related experimental results are also included in the chapter.

Research on the design of 3-D ICs has only recently begun to emerge. Many of the challenges remain unsolved, and significant effort is required to provide effective solutions to the issues encountered in the design of 3-D ICs. The major foci of this book are summarized in the last chapter and primary conclusions are drawn regarding further directions for research that will contribute to the maturation of this exciting solution to next-generation multifunctional systems-on-chip.

Chapter 2

Manufacturing of 3-D Packaged Systems

With the ongoing demand for greater functionality resulting in polylithic (multiple IC) products, longer off-chip interconnects plague the performance of microelectronic systems. The advent of the system-on-chip (SoC) in the mid-1990s primarily addressed the increasing delay of the off-chip interconnects. Integrating all of the components on a monolithic substrate enhances the overall speed of the system, while decreasing the power consumption. To assimilate disparate technologies, however, several difficulties must be surmounted to achieve high yield for the entire system while mitigating the greater noise coupling among the dissimilar blocks within the system. Additional system requirements for the radio frequency (RF) circuitry, passive elements, and discrete components, such as decoupling capacitors, which are not easily integrated due to performance degradation or size limitations, have escalated the need for technological innovations. Three-dimensional or vertical integration and system-on-package (SoP) have been proposed to overcome many of the inherent SoC constraints. The concepts of vertical integration and SoP are described in Sections 2.1 and 2.2, respectively, as well as the commonality between these innovative paradigms. Several three-dimensional technologies for 3-D packaging are reviewed in Section 2.3. Cost models related to these technologies are described in Section 2.4. The technological implications of 3-D integration at the package level are summarized in Section 2.5.

2.1 THREE-DIMENSIONAL INTEGRATION

One of the first initiatives demonstrating three-dimensional circuits was reported in 1981 [34]. This work involved the vertical integration of PMOS and NMOS devices with a single gate to create an inverter, considerably reducing the total area and capacitance of the inverter.

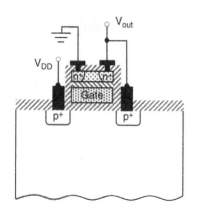

■ FIGURE 2-1 Three-dimensional stacked inverter [34].

A cross section of this inverter is illustrated in Figure 2-1. Several other approaches to three-dimensional integration, however, have been developed, both at the package and circuit levels. Bare or packaged die are vertically integrated, permitting a broad variety of interconnection strategies. Each of these vertical interconnection techniques has different advantages and disadvantages. Other more esoteric technologies for 3-D circuits have also been proposed. To avoid confusion, therefore, a crucial differentiation among the various 3-D integration approaches is maintained in this chapter. Two primary categories of 3-D systems are discerned, namely, system-in-package (SiP) and three-dimensional integrated circuits (3-D ICs). The criterion to distinguish between SiP and 3-D IC is the interconnection technology that provides communication for circuits located on different planes of a 3-D system. In SiP, through silicon vias (TSVs) with high aspect ratios are typically utilized. Due to the size of these vias, a high vertical interconnect density cannot be achieved. Hence, these interconnects provide coarse-grain connectivity among circuit blocks located on different planes. Alternatively, in 3-D ICs, fine-grain interconnection among devices on different planes is achieved by narrow and short TSVs.

2.1.1 **System-in-Package**

Henceforth in this book a system-in-package is described as an assemblage of either bare or packaged die along the third dimension, where the interconnections through the z-axis are primarily implemented through the following means:

- Wire bonding
- Vertical interconnects along the periphery of the die/package
- Long and wide, low-density vertical interconnects (in an array arranged across the die/package)
- Metallization between the faces of a 3-D stack

Die or package bonding can be implemented by utilizing a diverse collection of materials, such as epoxy and other polymers. Some examples of SiP structures are illustrated in Figure 2-2. Each example of these manufacturing techniques is discussed in Section 2.3.

2.1.2 **Three-Dimensional Integrated Circuits**

Three-dimensional IC manufacturing can be conceptualized either as a sequential or a parallel process. In the case of a sequential process, the devices and metal layers of the upper planes of the stack are grown on top of the first plane. Hence, the 3-D system can be treated as a

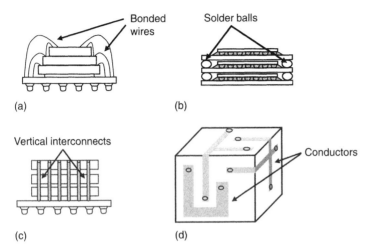

■**FIGURE** 2-2 Examples of SiP: (a) wire-bonded SiP [42], (b) solder balls at the perimeter of the planes [43], (c) area array vertical interconnects, and (d) interconnects on the faces of the SiP [44].

monolithic structure. Alternatively, a 3-D IC can be created by bonding multiple wafers or bare die. The distinctive difference between an SiP and a 3-D IC is the vertical interconnect structure. Communication among different die within a 3-D IC is implemented by

- High-density short and thin TSVs (placed at any point across the die not occupied by transistors)
- Capacitive coupling among parallel metal plates located on vertically adjacent die
- Inductive coupling among inductors located on vertically adjacent die

Some characteristic examples of these communication mechanisms for 3-D ICs are illustrated in Figure 2-3. Although 3-D ICs can be characterized as a subset of SiP, the advantages of 3-D vertical interconnects as compared to the interconnect mechanisms utilized in SiP are substantially greater. In addition, 3-D ICs face fewer performance limitations as compared with the SiP approach, unleashing the potential of three-dimensional integration. Various technologies for 3-D ICs are reviewed in Chapter 3.

2.2 **SYSTEM-ON-PACKAGE**

As portability has become a fundamental characteristic in many consumer products, systems with a small footprint, low weight, long battery life, and equipped with a variety of features typically attract a significant market share. Communications is the most remarkable feature of these products. Wireless communication, in turn, requires

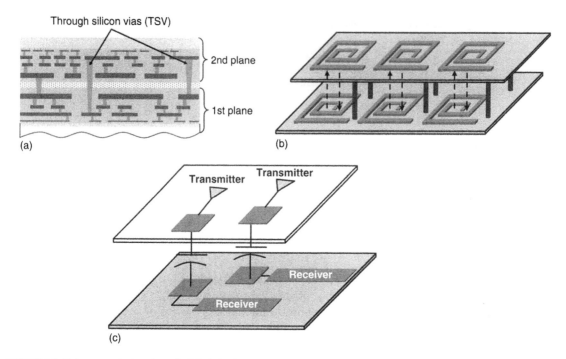

■**FIGURE 2-3** Various communication schemes for 3-D ICs: (a) short through silicon vias [45], (b) inductive coupling [36], and (c) capacitive coupling [46].

RF circuitry that includes passive components, such as resistors and high Q inductors that cannot be easily and cost effectively fabricated on the silicon. In addition, off-chip capacitors of different sizes are required to manage the voltage droop and simultaneous switching noise at the package level [47].

Implementing all of these components within a package and not as discrete components on a printed wiring board (PWB) contributes significantly to system miniaturization. System-on-package (SoP) is a system-level concept that includes passive components manufactured as thin films. High-performance transceivers and optical interconnects can also be embedded within a package [48]. State-of-the-art microvias provide communication among the embedded circuits and the mounted subsystems [49]. Microvias are vertical interconnects that connect the various components of an SoP through the laminated layers of the SoP substrate. Consequently, SoP supports wiring among the various subsystems as in an SiP, and can also contain system components integrated within the package.

A wide spectrum of subsystems that gather, process, and transmit data can be included in an SoP. Sensors and electromechanical devices that record environmental variables, analog circuits that process and convert the analog signals to digital data, digital circuits, such as application-specific ICs (ASICs), digital signal processing (DSP) blocks, and microprocessors, memory modules, and RF circuits that provide wireless communication make up a short and not exhaustive list of components that can be mounted on an SoP. Some of these subsystems can be vertically integrated, making an SoP a hybrid technology that combines planar ICs and 3-D circuits, as illustrated in Figure 2-4a. The set of functions that can be implemented by an SoP, SiP, 3-D IC, and SoC are qualitatively represented in Figure 2-4b. As expected, since an SoP is a system-integrating technology, an SoP can simultaneously include an SiP, 3-D IC, and SoC, providing significant functionality. An SiP is typically used in those applications that include memory modules and microprocessors or ASICs. 3-D ICs are currently more limited than an SiP (shown as a subset of an SiP in Figure 2-4b). Finally, SoCs can be manifested both as a planar system in an SoP and a three-dimensional component as an SiP or 3-D IC.

Several manufacturing and design challenges, as illustrated in Figure 2-5, need to be circumvented to ensure that the various forms of three-dimensional integration can be economically produced. From a technology perspective, a 3-D system should exhibit yield similar to a 2-D circuit. Since each plane within a stack can be separately prepared with a high-yield process, the primary challenge is to guarantee a high-yield bonding process. The bonding process should ensure reliable interconnectivity of the circuits located on different planes. In addition, the fabrication

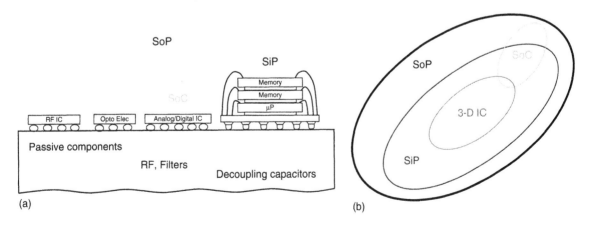

(a) (b)

■ **FIGURE 2-4** System-on-package is a hybrid design paradigm including both 2-D and 3-D ICs, (a) A typical SoP [48] and (b) relation of SoP, SiP, 3-D IC, and SoC.

■ **FIGURE 2-5** Manufacturing and design challenges for three-dimensional integration.

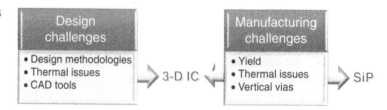

process of the vertical vias is a critical component contributing to the total system yield. Finally, advanced packaging technologies provide highly efficient heat transfer from the 3-D system to the ambient environment.

Overcoming these manufacturing challenges results in improved SiP in which reliable interconnect mechanisms that are more efficient than wire bonding can be employed, improving the performance of this form of three-dimensional integration. The potential of vertical integration, however, can be best exploited if various design challenges, as shown in Figure 2-5, are also met, resulting in general-purpose 3-D circuits. Design methodologies for 3-D circuits, however, barely exist. In addition, thermal management techniques for these high-density circuits are extremely important. The development of specific computer aided design (CAD) tools integrated into a design flow that can support the complexity of three dimensions is therefore an imperative need.

Mitigating these manufacturing and design issues will greatly promote the evolution of 3-D ICs, considerably expanding the application pool for this important manifestation of vertical integration. These technological developments will establish 3-D ICs as the primary representative of three-dimensional integration. Otherwise, vertical integration will primarily become an interconnection architecture (as in most SiPs) rather than an innovative design paradigm. An example of the predicted evolution of 3-D IC is illustrated in Figure 2-6.

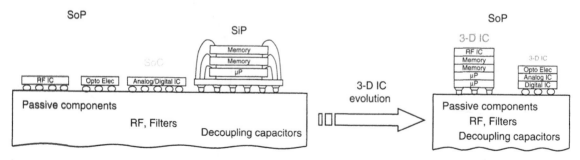

■ **FIGURE 2-6** System miniaturization through the integration of sophisticated 3-D ICs [48].

The development of a reliable package substrate that enables the embedding of both passive and active components is a challenging task. Minimum warpage and a low coefficient of thermal expansion (CTE) are basic requirements for package substrates in an SoP. Materials with a low CTE in the range of 8 to 12 ppm/°C have been developed [49]. In order to enhance the on-chip global interconnects, which are responsible for the longest on-chip delays, with low-loss off-chip wiring in the package, microstrip lines are fabricated with a 50 Ω impedance. Consequently, refined conductor geometries and low loss dielectrics are necessary [49], [50].

Dielectric materials play an important role in an SoP to provide both high performance and reliability. Dielectrics with high permittivity increase the amount of charge stored within a capacitor, while a low permittivity is required for those dielectrics that provide electrical isolation among neighboring interconnects, thereby reducing the coupling capacitance. A metric characterizing the loss of a dielectric material is the loss tangent (*tan δ*). This metric describes the ratio of the energy dissipated by the capacitor formed between the conductors and the energy stored in the capacitor. Low tangent loss for achieving low-latency signal transmission and high modulus for minimum warpage across the package is desirable. In addition, dielectrics should exhibit excellent adhesion properties to avoid delamination and to withstand, without any degradation in material properties, the highest temperature used to manufacture the package. Some of the dielectric materials that have been developed for SoP with the relevant material properties are listed in Table 2-1.

Table 2-1 High performance dielectric materials [49].

Dielectric Material	Dielectric Constant	Loss Tangent @ 1GHz	Modulus GPa	X,Y, CTE ppm/°C
Polyimide	2.9–3.5	0.002	9.8	3–20
BCB	2.9	<0.001	2.9	45–52
LCP	2.8	0.002	2.25	17
PPE	2.9	0.005	3.4	16
Poly-norbornene	2.6	0.001	0.5–1	83
Epoxy	3.5–4.0	0.02–0.03	1–5	40–70

2.3 **TECHNOLOGIES FOR SYSTEM-IN-PACKAGE**

A system-in-package constitutes a significant and widely commercialized variant of vertical integration. The driving force toward SiP is the increase in packaging efficiency, which is characterized by the die-to-package area ratio, the reduced package footprint, and the decreased weight. Homogeneous systems consisting of multiple memory dies and heterogeneous stacks including combinations of memory modules and an ASIC or a microprocessor are the most common type of SiP and have been commercialized by several companies [44], [51]. The typical communication mechanism among different planes of the stack is realized through wire bonding, as discussed in the following subsection.

2.3.1 **Wire-Bonded System-in-Package**

Wire bonding for a system-in-package is a commonplace technique. Due to the simplicity of the process, the required time to fabricate such an SiP is substantially shorter as compared to the turnaround time for an SiP that utilizes other interconnect techniques. Various wire-bonded SiP structures are depicted in Figure 2-7. The parameters that determine the packaging efficiency of these structures are the thickness of the bonded die and spacers, the thickness of the adhesive materials, the capability of providing multiple rows of wire bonding, and the size of the bumps used to mount the bottom die of the stack onto the package substrate. Alternatively, the performance of the stack is determined from the length of the bonding wires and the

■ **FIGURE 2-7** Wire-bonded SiP: (a) dissimilar dies with multiple row bonding, (b) wire-bonded stack delimited by spacer, (c) SiP with die-to-die and die-to-package wire bonding, and (d) top view of wire-bonded SiP [42].

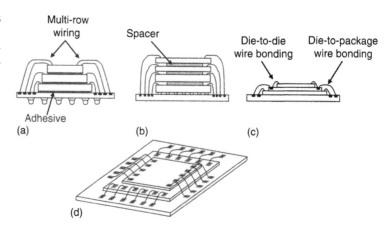

related parasitic impedances. Communication among the die of the SiP is primarily achieved through the package substrate and, as shown in Figure 2-7c, through wire bonding from one IC substrate to another IC substrate.

The manufacturing process for the structures shown in Figure 2-7 comprises the following key steps. All of the stacked dies or wafers are thinned with grinding and polished to a thickness of 50 µm to 75 µm [42]. Although aggressive wafer thinning is desirable, the die or wafer must be of sufficient thickness to avoid curling and cracking during the bonding process. The length of the bonding wires or, equivalently, the loop overhang, also depends upon the thickness of the attached wafers. In Table 2-2, the overhang requirement as a function of the die thickness is reported.

The bottom die of the stack is bumped and mounted in a flip chip manner onto the package substrate. The remaining die are successively adhered to the stack through thin-film epoxies. In certain cases, spacers between subsequent die are required. These spacers provide the necessary clearance for the bonding wire loop. Epoxy pastes and other thick-film materials are utilized with a thickness ranging from 100 µm to 250 µm [42]. A uniform gap between the attached die is required across the area of the spacer; therefore, spheres of spacer material can be utilized for this purpose. With these spacers, similar die or dies with quite diverse areas can be stacked on an SiP.

A wire-bonded SiP including up to four or five dies have been demonstrated [42], while two die stacking is quite common [52]. Increasing the number of die on an SiP is hampered by the parasitic impedance of the bonding wires and the available interconnect resources. For example, bonding wires can exhibit an inductance of a few nH. Consequently, a low loop profile is required to control the impedance characteristics of the wire. In addition, multiple rows for bonding can be utilized to increase the interconnect bandwidth. For multiple row wiring, low-height bonding wires are also necessary. The wire pitch for three bonding rows is listed in Table 2-3. Although the off-chip interconnects are decreased with wire bonding, limitations due to parasitic interconnect impedances and constraints on the number of die that can be wire bonded have led to other SiP approaches that provide greater packaging efficiency while exhibiting higher performance. These improved characteristics are enabled by replacing the bonding wires with shorter through-hole vias or solder balls [52]. This SiP technique is discussed in the following subsection.

Table 2-2 Loop overhang requirements vs. die thickness [42].

Die Thickness [µm]	Overhang [mm]
100	2.0
75	1.0
63	0.7

Table 2-3 Required bonding wire pitch for multiple row bonding [42].

Row No.	Wirebond Pitch [µm]
1	40
2	25 or 50
3	20 or 60

2.3.2 **Peripheral Vertical Interconnects**

To overcome the constraints of wire bonding, SiPs are implemented with peripheral interconnects formed with solder balls and through-hole vias. Several approaches based on this type of interconnect have been developed, some of which are illustrated in Figure 2-8 [44]. A greater number of die can be integrated with this approach as the constraints due to the wire parasitic impedance and the length are considerably relaxed. An interconnect structure utilized to provide a reduced pitch connection between the die and the package, known as an interposer, is included in each plane of an SiP, as shown in Figure 2-8. The interposer also provides the wiring between the die and the solder balls or through-hole vias at the periphery of the system. Several companies have developed similar SiP technologies. Micron Corporation uses the method shown in Figure 2-8b to produce high-density static random memory access (SRAM) and dynamic random memory access (DRAM) memory chips [53]. In another technique developed by Hitachi, vertical pillars are attached onto the printed circuit board (PCB), called PCB frames, to accommodate the vertical interconnects [54]. The I/Os of the IC are transferred through tape adhesive bonding (TAB) to the PCB frames, as illustrated in Figure 2-8c. High-density memory systems also use this method.

■ **FIGURE 2-8** SiP with peripheral connections, (a) solder balls [43], (b) through-hole via and spacers [53], and (c) through-hole via in a PCB frame structure [54].

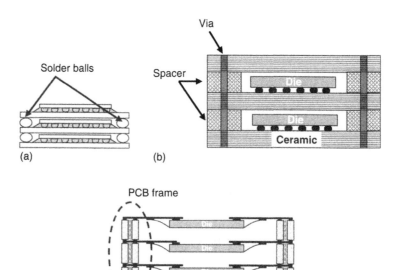

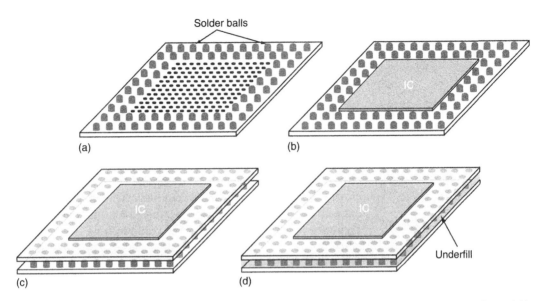

Solder balls

(a)   (b)

(c)   (d)

Underfill

■ **FIGURE 2-9** Basic manufacturing phases of an SiP: (a) interposer bumping and solder ball deposition, (b) die attachment, (c) plane stacking, and (d) epoxy underfill for enhanced reliability.

The basic steps of a typical process for SiP with peripheral interconnects are depicted in Figure 2-9 [43]. Initially, bumps are deposited on the interposers, while solder-plated polymer spheres are attached on the periphery of the interposers and provide communication among the dies of the stack. Next, the IC is attached to the interposer employing a flip chip technology. Vertical stacking of the interposers through the solder balls follows. The solder balls constitute an important element of the stacking process as these joints provide both the electrical connections and the mechanical support for the SiP. In addition, the height of the solder ball core provides the necessary space between successive interposers where the die is embedded. Solder balls with a height greater than 200 μm can therefore be utilized. If the solder balls collapse, the reliability of an SiP can be severely affected. Finally, in order to further reinforce the durability of this SiP structure, an epoxy underfill between adjacent interposers is applied. A similar technique has been developed by the Interuniversity Microelectronics Center (IMEC) in which the 3-D stack is achieved by directly soldering two contiguous solder balls ("ball on ball" bonding) [55].

Further enhancements of this bonding technique result in a decreased diameter of the vertical interconnects by utilizing through-hole electrodes and bumps with a diameter and depth ranging from 20 μm to

30 μm and 30 μm to 50 μm, respectively [56]. The through-hole electrodes are etched, and gold (Au) bumps are grown on the front and back side of the interposers to bond the IC. An important characteristic of this method is that only low temperatures are used to compress the stack. In addition, the thickness of the interposers is smaller than 50 μm, permitting greater packaging efficiency. The expected height of a ten-plane SiP implemented with this method is approximately 1 mm, a significant reduction in height over wire-bonding techniques.

Although through-hole vias yield greater packaging efficiency and mitigate certain performance issues related to wire bonding, a higher density vertical interconnect is required as interconnect demands increase. Indeed, with peripheral interconnects, the wire pitch decreases linearly with the number of interconnects. Alternatively, providing interconnects arranged as an array across the die of an SiP greatly increases wiring resources, while the interconnect pitch decreases with the square root of the number of interconnects [52].

2.3.3 **Area Array Vertical Interconnects**

An SiP with peripheral interconnects is inherently limited by the interconnect density and the number of components that can be integrated within a stack. Area array vertical interconnects improve the interconnect density among the planes of a 3-D system, offering opportunities for greater integration. Solder bumps or through-hole vias can be utilized for the area array interconnects; however, the processing steps for these two types of interconnects differ considerably.

In a manufacturing process developed by IBM and Irvine Sensors, the memory chips are bonded together to form a 3-D silicon cube [57], [58]. The wafers are processed to interconnect the I/O pads of the die to the bumps on the face of the cube. This metallurgical step is followed by wafer thinning, dicing, and die lamination to create the silicon cube. The stacking process is achieved through controlled thermocompression. An array of controlled collapse chip connect (C4) bumps provides communication among the planes of the cube. The interconnect density achieved with this technology is 1,400 C4 bumps, consuming an area of 260 mm^2. Note that with this technique, all of the planes within the cube share a common C4 bump array.

A higher contact density has recently been developed by 3D-PLUS and THALES [59]. A heterogeneous wafer is formed by attaching the individual ICs onto an adhesive tape that provides mechanical support. The I/O pads from each IC are redistributed with two different techniques into coarser pads whose size depends on the alignment accuracy prior to stacking the ICs. The first technique for pad redistribution utilizes two layers of copper embedded in benzocyclobutene (BCB). The average measured contact resistance is $\sim$2 Ω. Another technique for developing the vertical interconnects between the planes includes laminated films (as a dielectric) and copper. Despite the large number of steps, test vehicles have shown no signs of delamination among the adhered planes within the SiP.

Prior to bonding, the wafer thickness is reduced to 150 μm with grinding and mechanical polishing. Once the target thickness is achieved, the wafer is diced and tested for IC failures or open contacts during connection of the die I/Os with the coarser pads of the package. Only known good die (KGD) are therefore bonded, guaranteeing high yield. The 3-D system is connected through liquid adhesive and compression. The resulting SiP can contain up to six ICs, with a height less than a millimeter.

In these two techniques, each plane of the stack is designed and fabricated without any constraints imposed by the assembly process of the SiP, as the vertical interconnects are implemented on one face of the plane and not through the plane. Alternatively, the vertical interconnects are implemented with copper bumps and electrodes formed through the stacked circuits within the SiP [60]. The stacked die are aggressively thinned to a thickness of 50 μm, and copper (Cu) electrodes with a 20-μm pitch are formed. Square Cu bumps with a side length of 12 μm and a height of 5 μm are electroplated and covered with a thin layer of tin (Sn) (<1 μm) for improved bonding. The bonding is further enhanced by a nonconductive particle (NCP) paste, which is also used to encapsulate the Cu bumps.

Two bonding scenarios with different temperature profiles have been applied to stack four 50-μm thick dies on a 500-μm interposer [60]. In the first case, a bonding force is applied at high temperature (245°C), while in the second scenario, the temperature is reduced to 150°C accompanied by a thermal annealing step at 300°C. Both of these scenarios demonstrate good adhesion among the die, while the backside warpage is less than 3 μm for a square die with a side length of 10 mm.

With array-structured interconnects, the throughput among the die of the stack can be improved over an SiP with peripheral interconnect and bonding wires. The solder ball or bump size, however, constitutes an obstacle for vertically integrated systems with ultra-high interconnect density. Another concern is the rewiring that is necessary in some of these techniques, which prevents achieving the maximum possible reduction in the length of the interplane interconnects.

2.3.4 **Metallizing the Walls of an SiP**

Another SiP approach entails several ICs mounted on thin substrates with wiring or stacked PCBs to yield a 3-D system. Interconnects among the circuits on different dies are implemented through metal traces on the individual faces of the cube. The interconnect density achieved with this technique is similar to that of a wire-bonded SiP; however, a higher number of monolithic structures can be integrated within the cube.

One application of this type of SiP is a high density SRAM memory stack manufactured with a process invented at Irvine Sensors [61]. In this process, each module is separately tested before the stacking step. In this way, the KGD problem is significantly appeased as the faulty parts can be excluded or reworked in the initial steps. Fully operational ICs are selected and glued with a thin layer of adhesive material between each pair of ICs. The bonded stack undergoes a baking step with a specific curing profile. In order to reveal the buried metal leads from each memory module, the stack is plasma etched. If required, more than one face of the cube can be etched. The dimensions of the leads are 1 μm thick and 125 μm wide. The etching step exposes not only the metal leads but also the substrate of the stacked memory planes. Passivation is required to isolate the substrates from the metal traces created on the sidewall of the stack. Several layers of polyimide are deposited, where the thickness of each layer is greater than the exposed metal leads. Lift-off photolithography and sputter deposition are utilized to create the interconnects, including the busses, and the pads on the face of the cube with a titanium-tungsten/gold (Ti-W/Au) alloy. The pads on the sidewall of the cube connected to the metal leads of each IC form a T-shape interconnect, which is referred to as a T-connection. The resistance of a T-connection has been measured to be approximately 25 mΩ [61]. Finally, a soldering step is used to attach the stack onto a silicon substrate or PCB, permitting the stack to be connected to external circuitry.

Prototypes with as many as seven stacked ICs demonstrate a considerably higher integration as compared with a wire-bonded SiP. These prototypes are diced from an initial stack consisting of 70 identical ICs. Producing shorter stacks from a taller cube contributes to a reduction in the overall fabrication time. In addition to memory products, an image-capturing and processing system fabricated with a similar technique has been introduced [62], [63]. The entire fabrication process consists of several basic steps. Each IC is attached onto laminated films or, if necessary, a PCB along with discrete passive components. The substrate contains metal tracks, such as copper lines plated with gold, and the I/Os of the IC are wire bonded to these tracks. Each mounted IC is tested before bonding. This testing step improves the total yield as bonding the KGD is guaranteed. Testing includes validation of the interconnections connecting the IC to the laminated film or PCB. The test structures are placed in a plastic mold and encapsulated with an epoxy resin. After removal of the mold, a sawing step is applied to expose the metal tracks of each PCB within the stack. Alternatively, a deeper sawing can be applied to reveal the bonding wires rather than the metal tracks of the PCB, as shown in Figure 2-10. The cubical structure is plated with conventional electroplating techniques. The plated surfaces of the cube are patterned by a laser to produce the interconnects connecting the planes of the SiP. The typical dimensions for these sidewall interconnects are 6 μm to 7 μm thick and 500 μm wide. In the last step, the SiP can be soldered to a PCB on one face of the cube.

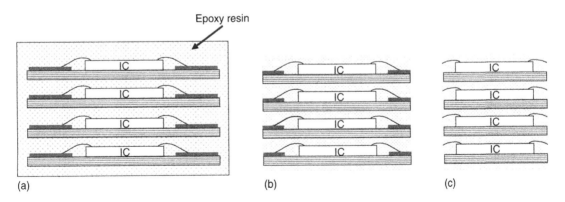

■ **FIGURE 2-10** Cross section of the SiP after removing the mold: (a) the SiP encapsulated in epoxy resin, (b) sawing to expose the metal traces, and (c) sawing to expose the bonding wires [64].

Although the methods used in each step of this technique are well known and relatively low cost, there are several design and efficiency concerns regarding the reliability of the surface contacts. These issues include the reliability of the internal contacts of each IC to the external interconnections on the cube faces, and the thermo-mechanical performance of a dense SiP. Finally, a particularly long interconnect path may be required to connect two circuits from different planes within an SiP. Consequently, utilizing thin and short vertical interconnects for communication among the planes within a 3-D system can result in higher performance. Different 3-D technologies result in dissimilar performance enhancements for an integrated system, while imposing a specific cost overhead due to the complexity of the fabrication process. A comparison of the manufacturing cost of the 3-D technologies, discussed in this chapter, is provided in the following section.

2.4 COST ISSUES FOR 3-D INTEGRATED SYSTEMS

The third dimension considerably enhances both circuit and packaging performance. A significant increase in speed and a decrease in power consumption are the primary improvements in circuit performance. In addition, the packaging density is greatly improved by efficiently exploiting the unused volume within the package. Alternatively, the fabrication cost of these technologies should be considered when selecting a target technology to satisfy the performance goals of an integrated system.

Two primary components of the manufacturing cost of an integrated circuit are the silicon area and the number of lithography steps required to fabricate a circuit. In 3-D systems, the footprint of the system is reduced. Multiplane structures, however, complicate the manufacturing process, primarily due to the need to bond the individual planes. The efficiency of the bonding process therefore directly affects the overall yield and varies according to the 3-D technology and the type of interplane communication, for example, wire bonds or through silicon vias. An analytic model that characterizes the cost of a 3-D system consisting of n planes is described by [65]

$$C = \frac{1}{Y_W^n Y_B^{n-1}} \frac{A_{design}}{A_{wafer}} [nM_W + (n-1)M_B], \tag{2-1}$$

where Y_W and Y_B are the yield of the process to fabricate each wafer and the yield of each bonding phase, respectively. Assuming wafer scale integration, A_{wafer} is the wafer area and A_{design} is the maximum footprint of the bonded planes. M_W and M_B are, respectively, the number of steps to fabricate each plane/wafer and the steps of the bonding process.

From this model, 3-D technologies decrease the cost of a system by reducing the physical footprint of the circuit. This improvement, however, is counterbalanced by the exponential decrease in the overall yield of the manufacturing process due to the multiple bonding and lithography steps. This situation suggests that the complexity of the 3-D manufacturing process can result in significantly higher cost for multiplane multifunctional systems as compared to a standard 2-D CMOS process typically used in systems-on-chip. A comparison of a multi-window display application and a graphics processor based on this model demonstrates that 3-D technologies produce a moderate increase of 49% and 43%, respectively, in the manufacturing cost [65]. The significant decrease in wirelength and, consequently, interconnect resources and the increase in the performance, however, justify the choice of a 3-D technology for high-performance applications.

The model of (2-1) is biased toward 2-D systems-on-chip since the manufacturing implications of combining different types of circuits, such as memory and logic, onto a single die such as an SoC are not considered. These yield limitations are removed in 3-D technologies, since each type of circuit can be placed on an individual plane and fabricated with the appropriate high-yield process. More accurate analytic models that consider these issues suggest that 3-D technologies can offer economic fabrication of multiplane systems for high-performance systems [66]. A theoretical comparison of a wireless sensor node and a third-generation mobile terminal implemented with different 3-D technologies is listed in Table 2-4. As listed in this table, when multiple aspects of a system, such as cost, performance, and temperature, are considered, 3-D integration can outperform planar systems-on-chip. Three-dimensional integration is therefore an excellent candidate for high-performance and heterogeneous systems. Furthermore, wafer scale integration of 3-D circuits with high-density vertical interconnects exhibits considerably higher performance as compared to a 3-D system-in-package. This attractive 3-D technology is discussed in the following chapter.

Table 2-4 Cost performance comparison of 3-D technologies [66].

Parameter	Wireless Sensor Node					3G Mobile Terminal				
	SoC	2-D SoP	3-D SiP	3-D wafer to wafer	3-D die to wafer	SoC	2-D SoP	3-D SiP	3-D wafer to wafer	3-D die to wafer
Normalized area	1.00	3.92	0.78	0.71	0.71	1.00	1.94	0.75	0.71	0.71
Overall yield	0.95	0.98	0.98	0.92	0.94	0.56	0.98	0.98	0.71	0.94
Normalized cost	1.00	4.11	4.04	1.14	2.96	1.00	0.40	0.40	0.38	0.33
Delay [ps]	127.37	176.36	148.33	83.90	83.90	317.88	205.37	168.34	259.63	259.63
ΔT (°C)	39.16	12.39	52.80	312.74	312.74	26.38	14.67	36.90	73.96	73.96

2.5 **SUMMARY**

The technological progress and related issues regarding three-dimensional integration at the package level can be summarized as follows:

- Three-dimensional packaging includes bare or packaged die vertically integrated using a variety of interconnection schemes.

- A system-in-package (SiP) and three-dimensional integrated circuits (3-D ICs) are currently the primary methods used for 3-D systems. A major difference between these approaches is the type of vertical interconnects used to connect the circuits located on different planes.

- A system-in-package is an assemblage of either bare or packaged die connected along the third dimension, where the interconnections through the z-axis are primarily implemented by wire bonding, metallization on the faces of the 3-D stack, vertical interconnects along the periphery of the die/package, and low-density vertical interconnects arranged as an array across the die/package.

- These different SiP interconnect structures are sorted in ascending order of vertical interconnect density.

- Three-dimensional ICs are interconnected with high-density short and thin TSVs, supporting low-level integration rather than die- or package-level integration as in an SiP.

- An SoP is a hybrid technology that combines planar ICs and 3-D circuits within a compact system. The primary difference as compared to an SiP is that an SoP supports more than wiring, containing system components within the package.

- An SiP has higher packaging efficiency and a smaller footprint and weight than a conventional 2-D system.

- Wire-bonded SiP are limited by the impedance of the bonding wires.

- SiP with peripheral interconnects are implemented by solder balls and through-hole vias. SiP are inherently limited by the interconnect density as the wire pitch decreases linearly with the number of interconnects.

- An SiP with array-structured interconnects can improve the interconnect density as compared to an SiP with peripheral interconnects and bonding wires as the interconnect pitch decreases with the square root of the I/O (or pin) requirements.

- Three-dimensional technologies increase the manufacturing cost of an integrated system. This additional manufacturing cost is counterbalanced by the decrease in the area, the increase in the die yield, and the increase in the performance of the overall system.

3-D Integrated Circuit Fabrication Technologies

A system-in-package (SiP) offers a large number of advantages over the traditional 2-D SoC, such as shorter off-chip interconnect lengths, increased packaging efficiency, and higher transistor density. These advantages provide significant performance improvements over SoC. Manufacturing issues, however, limit the scaling of the interchip interconnects, such as the wire bonds and solder balls, within an SiP. In addition, the inevitable increase in the delay of on-chip interconnects is not alleviated by SiP interconnect technologies. Although an SiP employing coarse grain through silicon vias can improve this delay, the lower density and impedance characteristics of these vertical interconnects limit the interplane interconnect bandwidth. The full potential of vertical integration is therefore not fully exploited. Consider, for instance, the various SiPs depicted in Figure 3-1, and assume that blocks A and B, located on different dies, are connected. The arrows represent typical interconnect paths connecting blocks A and B for each type of SiP shown in Figures 3-1a and 3-1b. A 3-D IC, shown in Figure 3-1c, provides the shortest interconnection between blocks A and B by utilizing through silicon vias. Consequently, 3-D ICs can achieve the greatest improvement in speed and power by decreasing the length of the long global interconnects as compared to other vertical integration technologies.

Three-dimensional ICs can be fabricated in either a batch (sequential) or parallel process. In the former case, the devices on the upper planes of a 3-D stack are grown on top of the first plane, resulting in a purely monolithic system. Fabrication processes based on this type of 3-D circuit structure are discussed in Section 3.1. Alternatively, some ICs are prepared separately before the bonding process and bonded to form a 3-D system. Such a 3-D IC is a polylithic structure and related manufacturing processes are described in Section 3.2. Other techniques

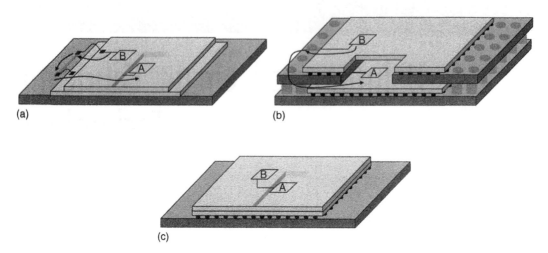

■FIGURE 3-1 Typical interconnect paths for (a) wire-bonded SiP, (b) SiP with solder balls, and (c) 3-D IC with through silicon vias (TSVs).

that provide contactless interplane communication have also been developed. Technologies for contactless 3-D ICs are reviewed in Section 3.3. One of the most important characteristics of polylithic 3-D ICs is the through silicon via, which provides electrical connection among the circuits on different planes, that is, the interplane interconnects. Due to the significance of these vertical interconnects, Section 3.4 is dedicated to discussing the manufacturing process of the vertical vias and related issues. The technological implications of 3-D integration are summarized in Section 3.5.

3.1 MONOLITHIC 3-D ICs

Monolithic 3-D ICs include two main types of circuits, stacked 3-D ICs and fin field effect transistors (fin-FETs). Stacked 3-D ICs include layers of planar devices successively grown on a conventional CMOS or silicon-on-insulator (SOI) plane. Alternatively, fin-FETs employ quasiplanar devices with certain advantages and limitations as compared to planar transistors. These two approaches are discussed in subsections 3.1.1 and 3.1.2, respectively.

3.1.1 Stacked 3-D ICs

The first 3-D ICs to be fabricated were stacked bulk CMOS or silicon-on-insulator (SOI) devices with simple logic circuitry that supported transistor-level integration. The devices that comprise a logic gate can be located on different layers and, more importantly, implemented with different technologies such as CMOS or SOI. Independent of the technology utilized for the first device layer, these transistors are fabricated with conventional and mature processes. For the devices on the upper planes, however, different fabrication methods are required. Several techniques, based on laser recrystallization or seed crystallization, are used to produce CMOS or SOI devices on the upper planes.

3.1.1.1 *Laser Crystallization*

After the work described in [67], several techniques using beam recrystallization have been developed to successfully fabricate 3-D ICs. In all of these techniques, the first device layer is fabricated with a traditional CMOS or SOI process. Note that only a device layer is fabricated on the first plane. The interconnect layers are not fabricated at this initial stage. Fabrication of the transistors on the upper planes should satisfy a twofold objective. The devices on the upper planes should exhibit satisfactory transistor electrical characteristics, such as field mobility, threshold voltages, and leakage currents. Alternatively, the characteristics of the transistors on the lower plane should not be degraded by the high temperatures incurred during the manufacturing process.

The first plane of devices is formed with a common MOS process. Depending on the fabrication process, this plane can include both [68] or only one type of device; either PMOS or NMOS [69]. In the latter case, the complementary type of device is fabricated exclusively on the upper planes. Alternatively, the upper planes can include only SOI devices [70].

Prior to the development of the upper device layer, an insulating layer of SiO_2 is deposited [68]. In order to protect the devices of the first plane from the elevated temperatures developed during the recrystallization phase, a thick layer with an approximate thickness of 1 micrometer is deposited on top of the insulating SiO_2 layer. This layer is a standard feature of each recrystallization technique and is composed of various materials, such as polysilicon [68], phosphosilicate glass (PSG) [69], and silicon nitride (Si_3N_4).

To grow the devices on the upper planes, polysilicon islands or thin polysilicon films are crystallized to single grains by an Argon laser [71]. The resulting 3-D IC based on this technique is shown in Figure 3-2, where the various layers are also indicated. The temperature

■ **FIGURE 3-2** Cross section of a stacked 3-D IC with a planarized heat shield (PHS) used to avoid degradation of the transistor characteristics on the first plane due to the temperature of the fabrication processes [68].

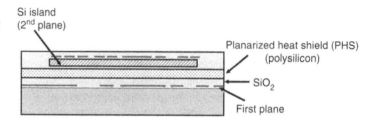

during the recrystallization phase can be as high as 950°C [70], while device formation with a lower temperature of approximately 600°C has been demonstrated [68].

Since the temperature used in these manufacturing steps approaches or exceeds the melting point of the metals commonly used for interconnects, doped semiconductor materials are used to interconnect the devices located on different planes. These materials can include n^+ doped polysilicon [68] or phosphorous-based polysilicon [70]. Alternatively, the interconnect layers can be fabricated after the upper plane devices are formed, where the contact holes are produced through reactive ion etching (RIE). The interplane interconnects are implemented by sputtering aluminum [69]. A cross section of a 3-D IC fabricated with aluminum interconnects is shown in Figure 3-3.

The major drawbacks of the laser recrystallization technique are the quality of the grown devices on the upper planes and the effect of the high temperatures on the electrical characteristics of the devices on the lowest plane. Comparison of the device characteristics within the two planes shows that the threshold voltage is sufficiently controlled, while the mobility of the devices on the upper plane is slightly worse than the transistors on the bottom plane [69]. Since the mobility of the devices on the upper planes is degraded, the

■ **FIGURE 3-3** Cross section of a stacked 3-D IC with a PMOS device on the bottom plane and an NMOS device in recrystallized silicon on the second plane [69].

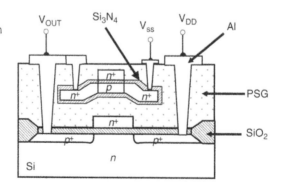

devices with lower mobility, namely, the PMOS transistors, are fabricated on the lower plane [35]. In addition, simple shift registers [68]
and ring oscillators [69] have been shown to operate correctly,
demonstrating that the devices on the first plane remain stable
despite the high temperature steps utilized to manufacture the devices
on the upper plane.

More advanced recrystallization techniques can provide up to three
device layers, where the lowest plane is a CMOS plane while the other
two planes are composed of SOI devices [70]. Since there are three
device planes in this structure, two isolation layers are required to
protect the devices on the lower planes. Measurements from sample
structures demonstrate that the threshold voltage of the devices is
well controlled with the devices on the upper planes exhibiting low
subthreshold currents [70]. The mobility of these transistors, however, can vary significantly.

Alternatively, E-beam has been used to recrystallize the polysilicon
to form the devices on the upper planes [71]. A three-plane prototype 3-D IC performing simple image processing functions has
been demonstrated [72]. On the topmost plane, consisting of
amorphous-Si (a-Si), the light is captured and converted into a digital
signal and stored on the bottom plane composed of the bulk-Si transistors. The intermediate plane, which includes the SOI transistors,
compares the digital data from one pixel to the digital data in adjacent
pixels to implement an edge detection operation. Measurements show
a narrow distribution of the threshold voltage of the devices on the
SOI plane with satisfactory *I-V* device characteristics. Although correct
operation of the circuitry is confirmed, demonstrating the capabilities
of a stacked 3-D IC, a large number of masks (34) is required, including five layers of polysilicon and two layers of aluminum interconnects. Although several recrystallization approaches have been
developed, none of these techniques has demonstrated the capability
of implementing complex, high-performance 3-D circuits, primarily
due to the inferior quality of the devices on the upper planes of the
3-D stack.

3.1.1.2 *Seed Crystallization*

Another technique used to fabricate multiple-device layers on bulk-Si
is crystallizing a-Si into polysilicon grains. Thin-film transistors (TFT)
are formed on these grains. The seed utilized for recrystallization can
be a metal, such as nickel (Ni), or another semiconductor, such as
germanium (Ge).

The basic processing steps of this technique are illustrated in Figure 3-4. A film of amorphous silicon is deposited with low-pressure chemical vapor deposition (LPCVD). A second film of low-temperature oxide (LTO), SiO$_2$, is deposited and patterned to form windows at the drain or both drain and source terminals of the devices for Ge or Ni seeding. Consequently, two kinds of TFTs are produced. One type is seeded only at one terminal, while the second type is seeded at both terminals. The deposition of the LTO SiO$_2$, patterning, and the deposition of the seed are additional steps as compared to a conventional process for TFTs. Thermal annealing is necessary to completely crystallize the channel films. Finally, the LTO and seeds are etched, permitting the TFTs to be fabricated with a standard process. A thermal layer of SiO$_2$ is used as the gate dielectric and boron and phosphorous doping is used for the junctions. The gate electrode is formed by *in situ* doped polysilicon, and the interconnections are implemented by modified interconnect

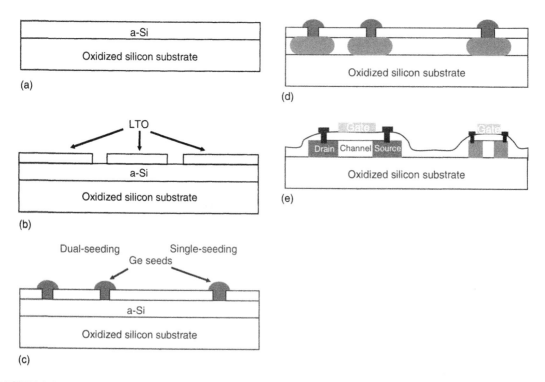

■ **FIGURE 3-4** Processing steps for laterally crystallized TFT based on Ge-seeding. (a) Deposition of amorphous silicon, (b) creating seeding windows, (c) deposition of seeding materials, (d) producing silicon islands, and (e) processing of TFTs [74].

and plug technologies from conventional 2-D circuits. The peak temperature of the process is 900°C.

Certain factors can degrade the quality of the fabricated TFTs, such as the size of the grains and the presence of defects within the channel region. Controlling the size and distribution of the grains contributes significantly to the quality of the manufactured devices. Comparisons among unseeded devices, produced by crystallizing a-Si, single-seeded, and dual-seeded TFTs, have shown that seeded devices exhibit enhanced device characteristics, such as higher field effect mobility and lower leakage currents. The greatest improvement in these characteristics is achieved by dual-seeded TFTs. The improvement in the characteristics of dual-seeded devices as compared to single-seeded transistors diminishes with device size. Thus, for small devices where the grain size of a single-seeded device is approximately the entire channel region, the performance of those TFTs is close to double-seeded TFTs as grain boundaries are unlikely to appear within the channel. Grain sizes over 80 μm have been fabricated [73]. High-performance TFTs formed by lateral crystallization on the upper planes exhibit low leakage currents and high mobility as compared to the SOI devices on the first plane, while significantly reducing the total area of the logic gates [73], [74].

Selective epitaxial growth (SEG) and epitaxial lateral overgrowth (ELO) have also been combined to fabricate multiple planes of SOI devices [75], [76]. The various steps of these techniques are summarized in Figure 3-5. The first device plane is formed on a thick layer of SiO_2. The oxide is patterned with photolithography and etching procedures to define the SOI islands (Figure 3-5a). A thinner SiO_2 layer, acting as the insulator, is deposited (Figure 3-5b). With photolithography, a window for SEG is opened within the silicon island patterns (Figure 3-5c). Using SEG and ELO, the patterns are filled by vertical and lateral growth of silicon within the SEG window (Figure 3-5d). The redundant silicon is etched with chemical mechanical planarization (CMP) (Figure 3-5e). A second layer of devices is fabricated with the same process, unlike the oxide deposition, which is achieved with plasma-enhanced chemical vapor deposition (PECVD). The transistors are fabricated on islands using conventional SOI techniques (see Figures 3-5f to 3-5h). The smaller island has an area of 150 nm x 150 nm, enabling an extremely high degree of integration on a single die. The photolithography and etching steps limit the dimensions of the SOI islands. Satisfactory device characteristics have been reported, with stacking faults only appearing on the first device layer [75].

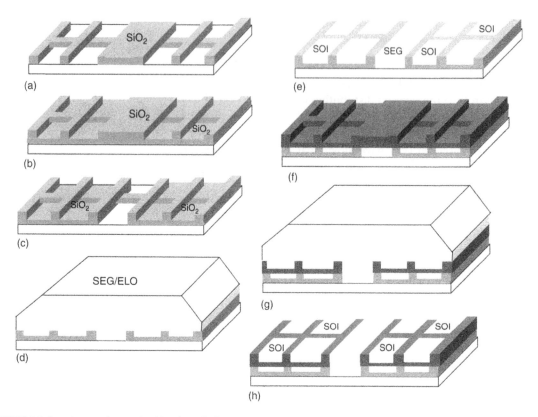

■ FIGURE 3-5 Processing steps for vertical and lateral growth of 3-D SOI devices [75].

3.1.1.3 Other Fabrication Techniques for Stacked 3-D ICs

Similar to 2-D circuits where standard cell libraries include compact physical layouts of individual logic gates or more complex logic circuits, standard cell libraries for 3-D circuits include "volumetric" logic cells with minimum volume and a smaller load as compared to 2-D cells. Fabrication techniques that use local clusters of devices to form standard 3-D cells have been developed [77] based on double-gate metal oxide semiconductor field effect transistors (MOSFETs) [78], [79]. The major fabrication stages for a 3-D inverter cell are illustrated in Figure 3-6, where a decrease of 45% in the total capacitance is achieved. Similar improvements are demonstrated for more complex circuits, such as a 128-bit adder, where a 42% area reduction is observed [77].

An SOI wafer is used as the first plane of the 3-D IC. The top silicon layer is thinned by thermal oxidation and oxide etching. A thin layer of oxide and a layer of nitride are deposited as shown in Figure 3-6a. The wafer is patterned, and shallow trench etching is used to define the active area (Figure 3-6b). A layer of LTO is deposited to fill the trench, which is planarized with CMP, where the nitride behaves as the stop layer (Figure 3-6c). The nitride film is removed, and LTO is utilized as a dummy gate. The LTO, silicon, and oxide are patterned and etched, followed by depositing a nitride film (Figure 3-6d). The drain and source regions for the first device layer are created with an As^+ implantation. As depicted in Figure 3-6e, a via is opened at the drain side to connect the terminals of the NMOS and PMOS transistors located on the second plane. The contact is implemented as a tunneling ohmic contact. A polysilicon layer is deposited and planarized by CMP utilizing the nitride film as the stop layer. The nitride film

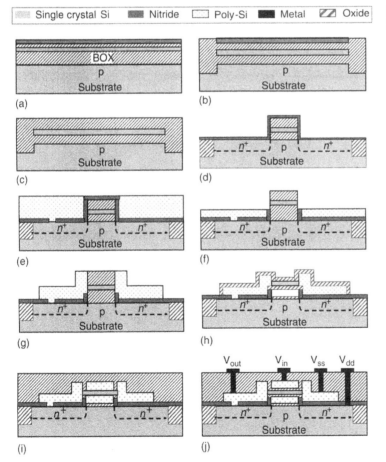

■ **FIGURE 3-6** Basic processing steps for a 3-D inverter utilizing the local clustering approach [77].

is etched to expose the silicon for the channel region (Figure 3-6f). Boron doping is used to implement the source and drain regions of the PMOS transistor on the top layer (Figure 3-6g). The active area for the devices is defined by removing the LTO and the buried oxide (Figure 3-6h). Before forming the gates for the stacked transistors, the gate oxide is grown by thermal oxidation, as shown in Figure 3-6i. Finally, *in situ* deposition of doped polysilicon forms the gate electrodes. Note that for the gate below the channel region of the PMOS transistor (see Figure 3-6j), a slow deposition rate is required to fully fill this region.

3.1.2 **3-D Fin-FETs**

3-D fin-FETs are based on quasiplanar fin-FETs [80] where the devices are stacked on top of each other, thereby sharing the same gate. A schematic of a 3-D fin-FET CMOS inverter is shown in Figure 3-7 [81], [82]. The second device layer can be either grown on top of the first plane or, alternatively, can be bonded from a second wafer. The most attractive feature of this technology is the approximately 50% reduction in gate area and routing resources to connect the devices within the gate. The drive current of the devices is not controlled by the width of the devices, W_n and W_p, but rather from the corresponding fin height notated as H_n and H_p for a NMOS and PMOS transistor, respectively, as shown in Figure 3-7.

The major steps of a 3-D fin-CMOS process are illustrated in Figure 3-8. A second oxide layer behaves as the buried oxide for the second plane. A silicon film is grown on top of the buried oxide layer. A mask for the gate, which can endure ion etching, is formed, and inductively plasma-enhanced (ICP) deep reactive ion etching (DRIE) is applied to

■**FIGURE 3-7** 3-D stacked fin-CMOS device [81].

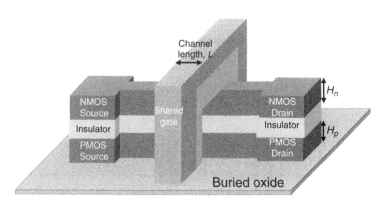

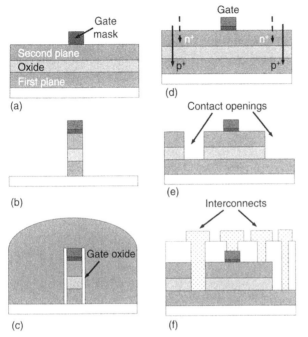

produce the fin structure, as shown in Figure 3-8b. The manufactured fins are almost vertical ($89.5°$–$90°$) [81]. A polysilicon film is deposited, surrounding the entire fin, and doped to form the gate electrode. Ions of boron and phosphorous in appropriate dosages are implanted to define the source and drain regions of both device layers. In this process, the NMOS transistors are located on the second plane. A portion of the boron ions is therefore present on the top device layer. The percentage of boron, however, is low—about 10% of the phosphorous concentration—resulting in negligible degradation in the device performance of the NMOS transistors. A critical fabrication factor is the contact between the drain regions of the top and bottom device layers. This connection is achieved with a surface-sidewall contact, slightly increasing the total active area of the devices on the bottom plane.

Fabricated devices include inverters, NAND gates, and static random memory access (SRAM) cells [81], [82], with satisfying characteristics, such as a high subthreshold slope and voltage transfer curves with sufficient noise margins. A disadvantage of this technology is that several additional design rules are required. Furthermore, stacking more than two planes with multiple metal interconnection layers is not straightforward; a common issue in all of the aforementioned techniques for stacked 3-D ICs. In addition, most 3-D manufacturing technologies are sequential which can therefore result in long manufacturing turnaround times. The requirement for high-quality devices for each plane of the 3-D system is an important limitation in stacked 3-D ICs. Furthermore, the insulator layers typically used as heat shields for several of these fabrication processes greatly impede the flow of heat during normal circuit operation, resulting in unacceptably high temperatures in the upper planes.

Alternatively, fabricating a 3-D system with vertically bonded ICs or wafers that are individually processed can reduce the total manufacturing time without sacrificing the quality of the devices on the upper planes of the stack. These techniques are discussed in the following sections.

3.2 3-D ICs WITH THROUGH SILICON (TSV) OR INTERPLANE VIAS

Wafer or die-level 3-D integration techniques, which utilize through silicon vias, are appealing candidates for 3-D circuits. Interplane vias offer the greatest possible reduction in wirelength with vertical integration. In addition, each plane of a 3-D system can be processed separately, decreasing the overall manufacturing time. As each IC of the 3-D stack is fabricated individually, the yield for each individual IC can be high. A broad spectrum of fabrication techniques for 3-D ICs has been developed. Although none of these techniques has been standardized, nearly all of these methods share similar fabrication stages, which are illustrated in Figure 3-9. The order of these stages, however, may not be the same, or some stages may not be used.

Initially, CMOS or SOI wafers are processed separately, producing the physical planes of the 3-D stack, while a certain amount of active area is reserved for the through silicon vias. The interplane vias are etched and filled with metal, such as tungsten (W) or copper (Cu), or even low-resistance polysilicon. In order to decrease the length of the through silicon vias, the wafers are attached to an auxiliary wafer, usually called a "handle wafer," and thinned to a different thickness depending on the technique. The alignment and bonding phase follows, as shown in Figure 3-9d. Finally, the handle wafer is removed from the thinned wafer, and the appropriate side of the wafer is processed and attached to another plane, if required. A broad gamut of materials and methods exist for each of these phases, some of which are listed in Table 3-1.

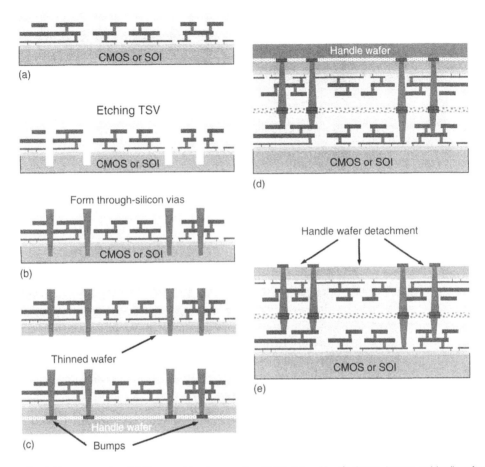

■ FIGURE 3-9 Typical fabrication steps for a 3-D IC process: (a) wafer preparation, (b) TSV etching, (c) wafer thinning, bumping, and handle wafer attachment, (d) wafer bonding, and (e) handle wafer removal.

Several of the proposed techniques for 3-D ICs support the integration of both CMOS and SOI circuits [86], [88]. From a fabrication point of view, however, SOI greatly facilitates the wafer-thinning step as the buried oxide (BOX) serves as a natural etching stop layer. This situation is due to the high selectivity of the etching solutions. Solutions with a Si to SiO_2 selectivity of 300:1 are possible [83]. In addition, SOI technology can yield particularly thin wafers or planes (<10 μm), resulting in short interplane vias. Alternatively, SOI circuits inherently suffer from poor thermal properties due to the low thermal conductivity of the oxide [90].

Table 3-1 Characteristics of fabrication techniques for 3-D ICs.

Process from	IC Technology	Interplane Vias Material	Wafer-Thinning Thickness	Alignment Accuracy [μm]	Plane Bonding Material	Bumps	Handle Wafer
[83], [84]	SOI	W	500 nm	±3	Cu-Cu pads	Yes	Yes
[45], [85]	SOI	Cu	∼10 μm	1-2	Polymers	No	No
[86]	SOI/CMOS	W	∼10 μm	N/A	Cu/Sn	Yes	Yes
[87]	CMOS	Cu		<2	Polymer	No	Yes
[88]	SOI/CMOS	n^+ poly –Si/W	7-35 μm	±1	Epoxy adhesive	Yes	Yes
[89]	SOI	Cu	<1	<1	Oxide fusion	No	Yes

The formation of the through silicon or interplane vias is an important issue in the design of high-performance 3-D ICs. This fabrication stage includes the opening of deep trenches through the interlayer dielectric (ILD) and the metal and device layers, passivation of the trench sidewalls from the metal layers and conductive substrate (for CMOS circuits), and filling the opening with a conductive material to electrically connect the planes of the 3-D system. Due to the importance of this fabrication stage, a more detailed discussion of interplane vias follows in Section 3.4.

Before stacking the plane, the wafers are typically thinned to decrease the overall height of the 3-D system and, therefore, the length of the interplane vias. A reduced wafer thickness, however, cannot sustain the mechanical stresses incurred during the handling and bonding phases of the 3-D process. The handle wafer is therefore attached to the original wafer prior to the thinning step. A handle wafer should possess several properties including the following:

- Mechanical durability to withstand the mechanical stresses incurred during the wafer-thinning process, due to grinding and polishing, and during bonding due to compressive and thermal forces.
- Thermal endurance to processing temperatures during wafer bonding and bumping.
- Chemical inertness to the solutions employed for wafer thinning and polishing.
- Simple and fast removal of the thinned wafer with appropriate solvents.
- Enhanced wafer alignment; for example, by being transparent to light.

Thinning can proceed with a variety of methods, such as grinding and etching, or both, accompanied by polishing. Various combinations can also be applied, such as a silicon wet etch followed by CMP, mechanical grinding with CMP, mechanical grinding succeeded by a spin etch, and dry chemical etching [85].

Accurate alignment of the thinned wafers is also a challenging task. A variety of techniques are utilized to align the wafer, based on the precise registration of the alignment marks. These mechanisms include infrared alignment, through wafer via holes, transparent substrates, wafer backside alignment, and intersubstrate alignment [91]. Typical alignment precisions range from 1 μm to 5 μm (see Table 3-1). Submicrometer accuracies have also been reported [89], [91].

Once the wafers are aligned, the bonding step follows. If die-to-wafer or die-to-die bonding is preferred, in order to avoid the integration of faulty dies, each wafer is diced prior to bonding and the dies are successively bonded. An increase in the turnaround time, however, is inevitable with this approach. Alternatively, the wafer can be diced and each die is separately tested. The known good dies (KGD) are placed on supporting wafers. Wafer-level bonding can be applied with high yield for each plane within a 3-D system (the overall yield can be affected by later processing steps, e.g., bonding). Face-to-face plane bonding is illustrated in Figure 3–9d. Alternatively, back-to-face bonding can be applied, resulting in slightly longer vertical interconnects between the planes. In addition, face-to-face bonding can be utilized if only two planes are integrated. A third plane is added in a back-to-face manner. Back-to-back bonding is also possible, resulting, however, in longer TSVs.

Bonding can be achieved with adhesive materials, metal-to-metal bonds, oxide fusion, and eutectic alloys deposited among the planes of the 3-D system [84]. Epoxies and polymers possess good adhesive properties and are widely used. Some of these polymers are listed in Table 3-2. Metal-to-metal bonding requires the growth of square bumps on both candidate planes for stacking and can consist of, for example, Cu, Cu/Ta, In/Au, and Cu/Sn. A portion of these bumps is deposited on the edge of the interplane vias, providing enhanced electrical connection between the planes. An epoxy filler can also be used to reinforce the bonding structure. An issue that arises in this case, however, is that the resin can result in bonding failure due to differences in the height of the metal bumps. To avoid such failures, pyramid- or conic-shaped bumps have been proposed [92].

Table 3-2 Several materials used for wafer bonding.

Polymer
Poly aryl ethel (FLARE) [93]
Methylsilesequionexane (MSSQ) [94]
Benzocyclobutene (BCB) [95]
Hydrogensilsesquioxane (HSQ) [96]
Parylene-N [97]

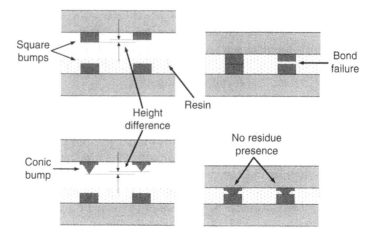

■ **FIGURE 3-10** Metal-to-metal bonding: (a) square bumps and (b) conic bumps for improved bonding quality [92].

The compressive bonding force causes the bumps to spread, excluding the resin from the bond region, as illustrated in Figure 3-10. Eutectic solders are also used for plane bonding where tin is plated on the plane surface followed by heating and compression [86]. In addition, the temperature profile depends on the material used for the interface and typically ranges from 200°C to 400°C so as not to degrade the copper inteconnections. Other highly refractory metals, such as tungsten or doped polysilicon, can be used if higher temperatures are necessary, increasing, however, the resistance of the interplane vias.

Depending on the bonding mechanism and material, certain requirements should be satisfied to avoid delamination and cracking of the planes:

- Small mismatch between the coefficient of thermal expansion (CTE) of the bonding material and the planes
- Mechanical endurance in the later processing steps
- Minimum wafer warpage
- No outgassing of the adhesives from heating, which can result in void formation
- No void generation due to the presence of residues on the layer surfaces
- Successful bonding can be verified by razor tests [45], where a razor blade is used to penetrate the interface between the planes, while other tests can include bending forces applied to the bonded wafer.

The final processing step for a 3-D IC with interplane vias is removing the handle wafer and cleaning the plane surface if subsequent bonding is required. A variety of solutions can be used to detach the handle wafer. The time required to accomplish this step also depends on the size of the wafer [86].

3.3 **CONTACTLESS 3-D ICs**

Although the majority of the processes that have been developed for 3-D ICs utilize vertical interconnects with some conductive material, other techniques that provide communication among circuits located on different planes through the coupling of electric or magnetic fields have been demonstrated. Capacitively coupled circuits are presented in subsection 3.3.1. Inductive vertical interconnects are discussed in subsection 3.3.2.

3.3.1 **Capacitively Coupled 3-D ICs**

In capacitively coupled signaling, the interplane vias are replaced with small on-chip parallel plate capacitors that provide interplane communication to the upper plane. A schematic of a capacitively coupled 3-D system is illustrated in Figure 3-11 [46]. A buffer is used to drive the capacitor. The receiver circuitry, however, is more complex, as the receiver must amplify the low-voltage signal to produce a full swing output. In addition, the receiver circuit should be sufficiently

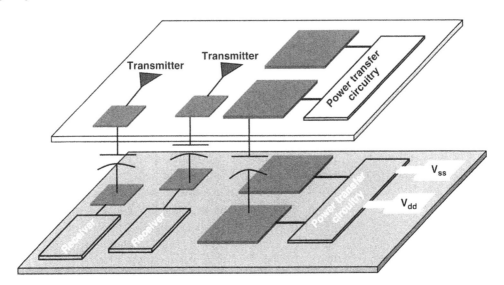

■ **FIGURE 3-11** Capacitively coupled 3-D IC; the large plate capacitors are utilized for power transfer, while the small plate capacitors are used for signal propagation [46].

sensitive and fast to detect and respond to the voltage being transferred through the coupling capacitors. The maximum voltage level that can be propagated through a coupling capacitance C_c is [98]

$$\Delta V_{max} = \frac{C_c}{C_c + C_p + \sum C_{tr}} \Delta V_{in},\tag{3-1}$$

where C_p is the capacitance of the capacitor plate to the substrate and adjacent interconnect preceding the receiver. The denominator in (3-1) includes the capacitance of the transistors at the input of the receiver.

A 5 fF capacitance has been used in [98] to transmit a signal to several receiver architectures. This capacitance is implemented with 20 μm × 20 μm electrodes with a separation of 2.5 μm (e.g., dielectric thickness) and a dielectric constant of 3.5. Simulations indicate correct operation for frequencies up to 500 MHz, while measurements show successful operation for signal transmission up to 25 MHz for a 0.5-μm CMOS technology. Smaller parallel plate capacitor structures and improved transceiver circuitry have also been reported, yielding a communication bandwidth of 1.23 Gb/sec [99]. For a 0.13-μm CMOS technology, the size of the capacitor electrodes is 8 μm × 8 μm. Comparing capacitively coupled 3-D ICs with SiP where the interplane interconnect is implemented by wiring over the chip edge (the total interconnect length is on the order of a centimeter), a 30% improvement in delay is demonstrated. In addition, a significant amount of dynamic power is saved with this interconnect scheme. The static power consumed by the receiving circuitry, however, decreases the overall power savings.

The parasitic capacitance of the devices is greatly reduced with a silicon-on-sapphire (SOS) technology, a type of SOI technology [46]. This situation is particularly advantageous for the power transfer circuitry realized with a charge pump. The transceivers and charge pump circuits have been fabricated in a 0.5-μm CMOS technology [46], utilizing capacitors with dimensions of 90 μm × 90 μm. The distance between the two planes is 10 μm. Decreasing the separation between the planes can increase the coupling, requiring smaller capacitor plates to propagate a signal. The capacitors used for the power exchange should be large, though, for enhanced coupling. A prototype based on this technique provides approximately 0.1 mA current to the devices on the upper planes. Successful operation up to 15 MHz has been demonstrated [46].

Critical factors affecting this technique are the size of the capacitors, which affects the interconnect density, the interplane distance, and the dielectric constant of the material between the planes, which determine the amount of coupling. Face-to-face bonding is preferable as the distance between the planes is less. Interconnecting a 3-D IC consisting of more than two planes with this approach, however, is a challenging task.

3.3.2 Inductively Coupled 3-D ICs

Inductive coupling can be an alternative for contactless communication in 3-D ICs. Each plane in a two-plane 3-D IC accommodates a spiral inductor, located at the same horizontal coordinates (see Figure 3-12). The size and diameter of the inductors are based on an on-chip differential transformer structure [100]. Inductors with a diameter of 100 μm are shown to be suitable for signal transmission, where the inductors are separated by a distance of 20 μm. In contrast to capacitively coupled 3-D ICs, the signal propagation is achieved through current pulses. As compared to the capacitive coupling technique, specialized circuitry is required for both the transmitter and receiver, dissipating greater power. A current-mode driver generates a differential signal. The receiver amplifies the transmitted current or voltage pulses, producing a full swing signal. Alternatively, inductive coupling does not require as small a separation of the planes as in the capacitive coupling technique, relaxing the demand for very thin spacing between planes. This advantage supports the integration of more than two planes for inductively coupled 3-D ICs. Simulations indicate a 5 Gb/sec throughput with 70 ps jitter at the output, consuming approximately 15 mW [100].

The main limitations of this technique relate to the size of the inductors, which yields a low-interconnect density as compared to the size of the interplane vias, increased power consumption of the transceiver circuitry, and interference among adjacent on-chip inductors. In addition, power delivery to the upper planes is realized with

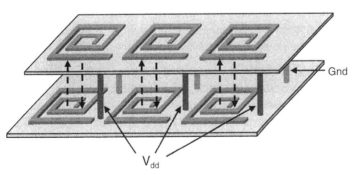

■ **FIGURE 3-12** Inductively coupled 3-D ICs. The galvanic connections are used for power delivery [100].

galvanic connections. Consequently, the fabrication process and cost for this type of 3-D IC is not greatly simplified as the via formation step is not eliminated.

3.4 VERTICAL INTERCONNECTS FOR 3-D ICs

Increasing the number of planes that can be integrated into a single 3-D system is a primary objective of three-dimensional integration. A 3-D system with high-density vertical interconnects is therefore indispensable. Vertical interconnects implemented as through silicon vias produce the highest interconnect bandwidth within a 3-D system, as compared to wire bonding, peripheral vertical interconnects, and solder ball arrays. Other important criteria should also be satisfied by the fabrication process for TSVs. A fabrication process for vertical interconnects should produce reliable and inexpensive TSVs. In addition, a TSV should exhibit low-impedance characteristics. A high TSV aspect ratio, the ratio of the diameter of the top edge to the length of the via, may also be required for certain types of 3-D circuits. The effect of forming the TSVs on the performance and reliability of the neighboring active devices should also be negligible.

As shown in Figure 3-9, TSVs are formed after the active devices on each plane of the 3-D circuit are fabricated and prior to the wafer-thinning step ("via first" approach). In this case, the TSVs are formed as blind vias exposed during the wafer-thinning step. This method provides the important advantages of compatibility with existent process flows and simplicity in wafer handling [101]. A disadvantage of this approach is the impact on reliability, resulting from wafer thinning and bonding. Alternatively, the TSVs are fabricated after the wafer-thinning step [102]–[104] ("via last" approach), as depicted in Figure 3-13. This approach alleviates those problems related to back-end processing; however, this method requires several processing steps with thin wafer handling, a potential source of manufacturing defects. Despite the disadvantages of the "via first" approach, via first is currently the most popular method for fabricating TSVs.

The technology for TSVs was greatly advanced by the invention of the BOSCH process in the mid-1990s, which was initially used to fabricate micro-electromechanical systems (MEMS) [105]. The BOSCH process essentially consists of two functions, namely, etching and deposition, applied in successive time intervals of different duration (typically on the order of seconds) [101], [106], [107]. Sulphurhexafluoride (SF_6) is utilized for the etching cycles, while fluorocarbon (C_4F_8) is used to

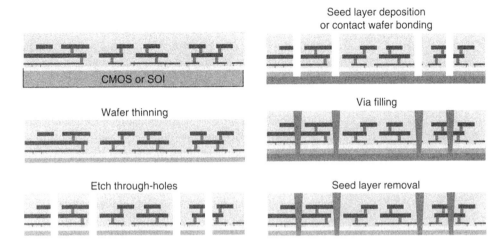

■ **FIGURE 3-13** Through silicon via formation and filling after wafer thinning ("via last" approach).

passivate the lateral wall of the TSV [105]–[107]. The BOSCH process is followed first by a barrier, after which a seed layer is deposited. The former layer, typically consisting of TiN or TaN deposition, prevents the Cu from diffusing into the silicon as in a conventional damascene process. The latter layer is utilized for the filling step of the TSV. Copper is used mostly for via filling. Although tungsten or low-resistance poly-Si can be used for the TSV, copper has the inherent advantage of compatibility with back end of the line (BEOL) processing and the multilayer interconnects used in modern ICs. Although the BOSCH process is effective in etching silicon, several issues regarding the quality of the TSVs have to be considered [101], such as

- Controllability of the via shape and tapering
- Conformal deposition and sufficient adhesion of the insulation, barrier, and seed layers
- Void-free filling with a conductive material
- Removal of excessive metal deposition along the edges of the TSV

TSVs have either a straight or tapered shape with various aspect ratios, as illustrated in Figure 3-14. The aspect ratio (*i.e.*, D/W) ultimately depends on the thickness of the wafer that results from the wafer-thinning step, where this ratio can range from 5 to 15. TSVs with a wide range of diameter values have also been manufactured [101], [102], [105]–[110]. Tapered vias are preferred to nontapered vias, as a lateral wall with a slope and smooth surface facilitates the deposition of the barrier and seed layers and the following filling step

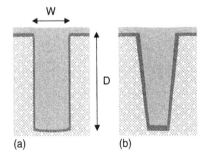

■ **FIGURE 3-14** Through silicon via shapes: (a) straight and (b) tapered.

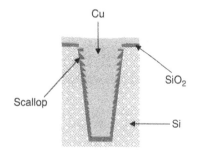

Cu

SiO₂

Scallop

Si

■ FIGURE 3-15 A schematic representation of the scallops formed due to the time-multiplexed nature of the BOSCH process.

[101]. This behavior occurs because the tapered profile decreases the effective aspect ratio of the straight segment of the TSV. Excessive tapering, however, can be problematic, leading to V-shaped vias at the bottom edge. In addition, a specific interconnect pitch is required at the bottom side of the via; consequently, tapering should be carefully controlled.

Although the BOSCH process can etch silicon sufficiently fast, producing a smooth surface, barrier and seed layer deposition is not a straightforward task. The BOSCH process results in a particularly rough surface, producing scallops, as illustrated in Figure 3-15 [109]. Note that the roughness of the surface decreases with the via depth. The rough sidewall is due to the time multiplexed cycles of the etching and deposition steps of the BOSCH process. A low surface roughness is of particular importance for those steps that follow the via formation process. The scallops not only prevent a conformal barrier and seed layer deposition but also increase the diffusion of the copper into the silicon despite the presence of the barrier layer [108].

Various thermal loads can result from the following processing steps, such as wafer bonding, where thermocompression is typically utilized. Finite element analysis of TSVs with a diameter ranging from 3 μm to 10 μm has demonstrated that thermal loads contribute more to the induced stresses on a TSV as compared to those caused by the bonding force. For specific bonding conditions where the applied bonding force is 300 N and the temperature ranges from 300°C to 400°C, the thermal loads constitute 84% of the total induced stress [111]. Simulation of the thermal loads imposed on the TSVs has indicated that the thermal stress, due to CTE mismatch between the dielectric and metal filling, are more pronounced at the region where the scallops are sharper [109]. These stresses can crack the dielectric layer, increase the current leaking into the silicon substrate, or cause delamination of the copper within the TSV.

Another problem related to deep reactive ion etching concerns the silicon undercut below the mask at the upper and wide edge of the TSV. Uncontrolled undercut can lead to mask overhang, which in turn can accelerate plating at the top of the TSV. Faster plating of the via upper edge can lead to premature closing of the via and to formation of a void inside the TSV [106].

Two key parameters can be used to adjust the surface roughness of the TSV sidewalls; the ratio of the time duration of the etching and passivation steps, and the flow of the C_4F_8 [107]. In addition, variations of the BOSCH process or two-step processes can be used to adjust the roughness of the TSV surface. For example, manufacturing process parameters, such as the pressure, bias power, and etching cycle time, can be varied in time rather than maintained static throughout the etching step. With this approach, the roughness of the surface has been decreased from ~ 0.05 μm to ~ 0.01 μm [109].

Via formation is followed by deposition of the barrier and seed layers. The primary goal of this step is to achieve a conformal profile throughout the TSV depth, which is increasingly difficult for TSVs with high aspect ratios. Different chemical vapor depositions are typically utilized for this step in order to provide good uniformity with moderate processing temperatures. Metalorganic CVD (MOCVD) can be utilized to deposit the barrier layer. The disadvantage of this method is poor adhesion. Alternatively, physical vapor deposition (PVD) can be applied for layer deposition, which yields fair adhesion at low temperature. PVD, however, results in poor uniformity.

Via filling is achieved by electroplating, where the copper filling is grown laterally on the deposited seed layer. Alternatively, electroplating can be implemented in a bottom-up manner where a contact wafer is attached to the bottom of the device wafer, which includes the via openings. A major problem related to this technique is the difficulty in removing the contact wafer, which provides the seed layer for the via filling [107]. The copper electroplating should produce uniform and void-free TSVs to achieve high-quality signal paths. Poor electroplating conditions that result in void formation within the TSV are illustrated in Figure 3-16. The requirement for void-free via filling can be achieved by maintaining a constant deposition rate throughout the via depth. Due to the tapered shape of the TSV, however, achieving a constant deposition rate requires the continuous adjustment of various parameters of the electroplating process. Parameters that affect the TSV filling profile, which can be altered during copper deposition, include the solution composition, wafer rotation speed, applied current waveform, current pulse duration, and current density [101], [106], [107]. Current waveforms, such as direct current (DC), forward pulse current (PC), and reverse pulse current (PR), can be applied. Since the deposition rate on the top edge of the TSV is larger than at the bottom, a low forward current density and a high reverse current density are used to maintain a constant deposition rate [101], [102], [107]. As filling at the bottom edge progresses, thereby closing the via opening, the current density is appropriately adjusted to maintain a fixed deposition rate.

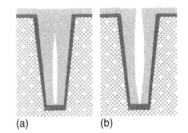

(a)   (b)

■ **FIGURE 3-16** Schematic representation of poor TSV filling resulting in void formation (a) large void at the bottom and (b) seam void.

■ **FIGURE 3-17** Structure of partial TSV and related materials [112].

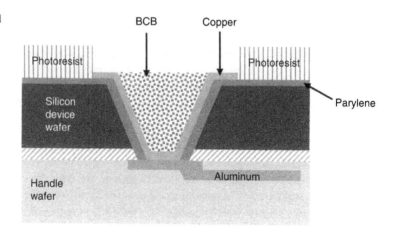

■ **FIGURE 3-17** Structure of partial TSV and related materials [112].

Another alternative to avoid via filling problems is to utilize partially filled vias [112]. A partially filled via is shown in Figure 3-17, where the various layers of the structure are also illustrated. The tapering of these TSVs ranges from 75° to 80°. The nonfilled metal volume of the etched via hole is filled with benzocyclobutene (BCB), while a layer of parylene insulates the silicon substrate from the TSVs. A concern regarding these partially filled TSVs is the electrical resistance due to the reduced amount of plated metal. Experimental results listed in Table 3-3, however, demonstrate that low-resistance vias can be achieved. Another important issue related to this type of TSV is that the density of metal within the volume of a 3-D circuit is reduced, while the density of the dielectric (which fills the etched via openings) is increased. This situation increases the thermal resistance in the vertical direction, which is the primary direction of the heat flow, requiring a more aggressive thermal management policy and possibly reducing the reliability of the circuits.

Table 3-3 Resistance of partially filled TSV [112].

Via Bottom Diameter [μm]	Via Top Diameter [μm]	Resistance of one TSV [mΩ]
60	100	30
80	120	28
100	140	24

Any excessive metal concentrated on the top edge of the TSV should be removed prior to wafer bonding. This metal removal is typically implemented by CMP. A copper annealing step can also be present either before or after the CMP step [100]. After the thick metal residue is removed from the wafer surface, wafer thinning and bonding follow, preceded, if necessary, by a rewiring step. In addition to manufacturing reliable TSVs, TSVs should exhibit low-impedance characteristics, as discussed in the following section.

3.4.1 **Electrical Characteristics of Through Silicon Vias**

The electrical characteristics of a TSV are of primary importance in 3-D ICs. A variety of TSVs with significantly different aspect ratios and electrical parameters have been reported [101], [102], [106]–[110]. Some relevant examples are listed in Table 3-4. The resistance values include, if appropriate, the resistance of the contact on which the TSV is placed.

Electrical characterization of these interconnect structures is a crucial requirement, since electrical models are necessary to accurately describe the interconnect power and speed of a 3-D circuit. An electrical model of a TSV is illustrated in Figure 3-18, where the vias are modeled as an *RLC* interconnect. This model is based on fitting measurements of the S-parameters with the parameters of the model. The test vehicle is a ground-signal-ground TSV structure connected to a coplanar waveguide. The height and diameter of the TSVs are 150 µm and 50 µm, respectively. The value of the model impedances are listed in Table 3-5. The resistance and inductance of the TSV structure shown in Figure 3-18 are empirically described, respectively, by [114]

Table 3-4 Dimensions and electrical characteristics of the through silicon vias.

Process from	Depth [µm]	Diameter [µm]	Total Resistance [mΩ]
[87]	25	4	140
[113]	30	2×12	230
[110]	80	5/15	9.4/2.6
[110]	150	5/15	2.7/1.9
[114]	90	75	2.4

$$L_{TSV} = \frac{L_0}{1 + \log(f/10^8)^{0.26}},\qquad(3\text{-}2)$$

$$R_{TSV} = R_o\sqrt{1 + \frac{f}{10^8}},\qquad(3\text{-}3)$$

where the skin effect is considered. f is the maximum signal frequency propagated through the TSV. Comparison of these expressions with measured S-parameters [114] exhibit negligible error for frequencies up to 20 GHz.

Table 3-5 Parameters of the electrical TSV model shown in Figure 3-18 [114].

Parameter	Definition	Value
C_{ox_TSV}	Capacitance between TSV and thin oxide layer	910 fF
C_{Si}	Capacitance of the silicon substrate	9 fF
C_{ox}	Capacitance of SiO_2 on the silicon surface and fringing capacitance between vias	3 fF
G_{Si}	Resistive losses between signal and ground TSV through the silicon substrate	1.69 m/Ω
L_0	TSV inductance at 0.1 GHz	35 pH
R_0	TSV resistance at 0.1 GHz	12 mΩ

■ **FIGURE 3-18** Electrical model of a TSV [114].

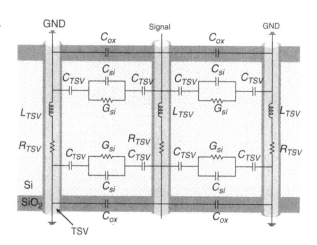

3.5 **SUMMARY**

Various manufacturing technologies and related issues for 3-D ICs can be summarized as follows:

- 3-D ICs can be realized with either a sequential or parallel manufacturing process. Sequential processes result in monolithic structures, while parallel processes yield polylithic structures.

- Stacked 3-D ICs are monolithic structures that significantly reduce the total gate area and capacitance. Stacked 3-D ICs can be fabricated by laser or e-beam recrystallization and silicon growth based on semiconductor or metal seeding.

- 3-D Fin-Fets share a fin-shaped gate that can reduce the area of the logic gates; however, the design complexity increases and cannot be easily extended to 3-D ICs with more than two planes.

- Wafer or die-level 3-D integration techniques, which utilize through silicon vias, are an appealing candidate for 3-D circuits as these vias offer the greatest reduction in wirelength.

- Fabricating 3-D ICs with TSVs typically includes the following steps: wafer preparation, TSV formation, wafer thinning, bumping, handle wafer attachment, wafer bonding, and handle wafer removal.

- Plane bonding can result from adhesive materials, metal-to-metal bonds, oxide fusion, and eutectic alloys.

- Coupling of electric or magnetic fields can be used to communicate among circuits located on different planes, producing contactless 3-D ICs.

- Some limitations of contactless 3-D ICs are the size of the inductors and capacitors, small distance between planes, and power delivery to the upper planes.

- The formation of a TSV is largely based on the BOSCH process, which consists of etching and deposition steps, multiplexed in time. Electroplating is the most common technique for TSV filling.

- Related problems with the BOSCH and electroplating techniques are rough surfaces, mask undercut, and void formation.

- To produce reliable TSVs, certain parameters of the BOSCH and electroplating techniques should be dynamically adjusted. These parameters include the solution composition, wafer rotation speed, applied current waveform, current pulse duration, and current density.

Chapter 4

Interconnect Prediction Models

In deep submicrometer technologies, the interconnect resistance has increased significantly, profoundly affecting the signal propagation characteristics across an IC. On-chip interconnects have, therefore, become the primary focus in modern circuit design. Considering the broad gamut of interconnect issues, such as speed, power consumption, and signal integrity, beginning the interconnect design process early in the overall design cycle can prevent expensive iterations at later design stages, which increase both the design turnaround time and the cost of developing a circuit. Early and accurate estimates of interconnect-related parameters, such as the required metal resources, maximum and average wirelength, and wirelength distribution are, therefore, indispensable for deciding on the nature of the interconnect architecture within a circuit.

Interconnect prediction models aim at providing a distribution of the length of the nets within a circuit without any prior knowledge of the physical design of the circuit. This *a priori* distribution can be utilized in a general interconnect prediction framework, producing estimates for various design objectives. Interconnect prediction models have existed for a long time. Recently, these models have evolved to consider interconnect issues, such as placement and routing for interconnect-dominated circuits, interconnect delay, and crosstalk noise. The majority of these models is based on either empirical heuristics or rules [115], [116]. A well-known rule, the Rent's rule, forms the basis for the majority of interconnect prediction models. This rule and a related interconnect prediction model for 2-D circuits are discussed in the following section. Several interconnect prediction models adapted for 3-D circuits are presented in Section 4.2. Projections for 3-D circuits obtained from these models that demonstrate the opportunities of vertical integration are discussed in Section 4.3. The key points of this chapter are summarized in Section 4.4.

4.1 INTERCONNECT PREDICTION MODELS FOR 2-D CIRCUITS

The cornerstone for the vast majority of wirelength prediction models is an empirical expression known as Rent's rule developed in the early 1960s [115]. Although the derivation of this rule was based on partitions of circuits consisting of only tens of thousands of gates, the applicability of this rule has been demonstrated for numerous and more complex circuits containing millions of gates [117]–[119]. Rent's rule is described by a simple expression,

$$T = kN^p, \tag{4-1}$$

correlating the number of I/O terminals of a circuit block with the number of circuit elements that are contained in this block. T and N are the number of terminals and elements of a circuit block, respectively. k is the average number of terminals per circuit element, and p is an empirical exponent. A circuit element is an abstraction and does not refer to a specific circuit; consequently, a circuit element can vary from a simple logic gate to a highly complex circuit. In addition, each circuit block has a limited number of terminals and a specific circuit element capacity. The type of circuit and partition algorithm determine the specific value of k and p [115]. Although the parameter p has initially been determined to be in the range of $0.57 \leq p \leq 0.75$, higher values for p have also been observed. In general, the parameter p increases as the level of parallelism of a system increases [116], [120].

Determining the interconnect length distribution is equivalent to enumerating the number of connections $i_{2-D}(l)$ for each possible interconnect length l within a circuit. Such an enumeration, however, is a formidable task. To circumvent this difficulty, a less computationally expensive approach can be utilized [121], [122]. Beginning with an infinite homogeneous sea of gates as illustrated in Figure 4-1, the number of interconnects within an IC with length l is determined as

$$i_{2-D}(l) = M_{2-D}(l)I_{gp}(l), \tag{4-2}$$

where $I_{gp}(l)$ is the expected number of connected gate pairs at a distance l in gate pitches and $M_{2-D}(l)$ is the number of gates producing these pairs. In order to obtain the number of gate pairs connected with an interconnect of length l, Rent's rule is employed where a circuit element is considered to be equivalent to a simple gate. In addition, the total number of terminals, including both the I/O terminals and the terminals connecting different circuit blocks for any partition of the original circuit, remains the same. Furthermore, Rent's rule can be applied to any partitioning algorithm, since a different partitioning technique alters the empirical exponents included in Rent's rule, rather than the expression describing this rule. Consequently, the expected number of interconnects between two groups of gates N_A and N_C located at a distance l is

$$I_{gp}(l) = \frac{ak}{N_A N_c}\left((N_A + N_B)^p + (N_B + N_C)^p - (N_B)^p - (N_A + N_B + N_C)^p\right),$$

$$(4\text{-}3)$$

where N_B is a group of gates within a distance l from group N_A. To approximate the number of interconnects between two groups of gates N_A and N_C separated by l gate pitches, as shown in Figure 4-1, the following observation is used.

For a sufficiently large number of gates N, those gates that comprise group N_C form a partial manhattan circle (the manhattan distance is the metric used to evaluate the distance among the gates or, equivalently, the interconnect length among the gates). This situation applies to most of the gates within a circuit except for those gates close to the periphery of the IC. Consequently, from Figure 4-1,

$$N_A = 1, \qquad\qquad (4\text{-}4)$$

$$N_B(l) = \sum_{k=1}^{l-1} 2k = l(l-1), \qquad\qquad (4\text{-}5)$$

$$N_C(l) = 2l, \qquad\qquad (4\text{-}6)$$

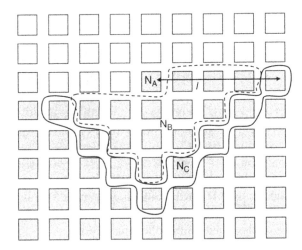

■ **FIGURE 4-1** An example of the method used to determine the distribution of the interconnect length. Group N_A includes one gate, group N_B includes the gates located at a distance smaller than l (encircled by the dashed curve), and N_C is the group of gates at distance l from group N_A (encircled by the solid curve). In this example, $l = 4$ (the distance is measured in gate pitches) [121].

where N_B corresponds to the number of gates encircled within the partial manhattan circle formed by the gates in group N_C. The gates included in N_C shape a partial but not full manhattan circle because the crosshatched gates are not considered to be part of N_C. This elimination of gates during the enumeration process ensures that a gate is not counted more than one time. With these assumptions and by employing a binomial expansion, an approximate expression for $I_{gp}(l)$ is obtained,

$$I_{gp}(l) \cong ak\frac{p}{2}(2-2p)l^{2(p-2)}. \qquad (4\text{-}7)$$

From (4-7), the number of gate pairs connected by an interconnection of length l is a positive and decreasing function of length for $p < 1$. The factor a in (4-7) considers the number of terminals acting as sink terminals and is

$$a = \frac{f.o.}{1+f.o.}, \qquad (4\text{-}8)$$

where $f.o.$ is the average fanout of a gate. By incorporating the number of gate pairs $M_{2\text{-}D}(l)$, which can be determined by simple enumeration, and using the binomial approximation for $I_{gp}(l)$, the interconnect length distribution $i_{2\text{-}D}(l)$ can be obtained as

$$i_{2-D}(l) = \frac{ak}{2}\Gamma\left(\frac{l^3}{3} - 2\sqrt{N}l^2 + 2Nl\right)l^{2(p-2)} \ for \ 1 \leq l\sqrt{N}, \qquad (4\text{-}9)$$

$$i_{2-D}(l) = \frac{ak}{6}\Gamma\left(2\sqrt{N}-l\right)^3 l^{2(p-2)} \ for \sqrt{N} \leq l < 2\sqrt{N}, \qquad (4\text{-}10)$$

where Γ is a normalization factor. The cumulative interconnect density function (c.i.d.f.) that provides the number of interconnects with a length smaller or equal to l is

$$I_{2-D}(l) = \int_1^l i_{2-D}(z)dz. \qquad (4\text{-}11)$$

Expressions (4-9)–(4-11) provide the number and length distribution of those interconnects within a homogeneous (*i.e.*, circuit blocks with the same capacity and number of pins) 2-D circuit consisting of N gates. By considering more than one device plane where each plane is populated with identical circuit blocks, the wirelength distribution for a homogeneous 3-D circuit can be generated. Various approaches have been followed to determine the wirelength distribution of a 3-D circuit, as described in the following section.

4.2 **INTERCONNECT PREDICTION MODELS FOR 3-D ICs**

Several interconnect prediction models for 3-D ICs have been developed, providing early estimates of the wirelength distribution in a 3-D system. Although these kinds of models lack any knowledge about the physical design characteristics of a circuit, such models can be a useful tool in the early stages of the design flow to roughly predict various characteristics of the system, such as the maximum clock frequency, chip area, required number of metal layers, and power consumption. Specifically, for 3-D ICs, such models are of considerable importance because a mature commercial process technology for 3-D ICs does not yet exist. Interconnect prediction models are, therefore, an effective means to estimate the opportunities of this innovative and developing technology.

A wirelength prediction model for 3-D circuits should consider not only the interconnections among gates located within the same plane, but also the interconnections among gates on different device planes. Consequently, in 3-D circuits, the partial manhattan circles formed by the gates in N_C as illustrated in Figure 4-1 become partial manhattan spheres as gate pairs exist not only within the same plane (intraplane gate pairs) but also among gates located on different planes (interplane gate pairs). These spheres are formed by partial manhattan circles of reduced radius, which enclose the interplane gate pairs. A decreasing radius is utilized for the interplane gate pairs to consider the vertical distance among the physical planes. This decreasing radius produces a spherical region rather than a cylindrical region for enumerating the interplane gate pairs. Such a hemisphere is illustrated in Figure 4-2, where the vertical distance between

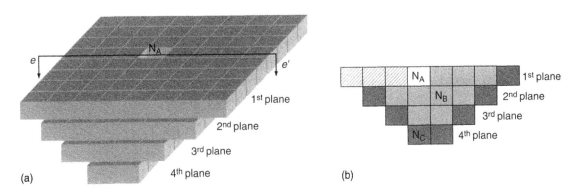

■ FIGURE 4-2 An example of the method used to determine the interconnect length distribution in 3-D circuits, (a) partial manhattan hemisphere and (b) cross section of the partial manhattan hemisphere along *e-e′*. The gates in N_B and N_C are shown with light and dark gray tones, respectively [126].

adjacent planes is assumed to be equal to one gate pitch. Considering different lengths for the vertical pitch results in hemispheres with different radii.

The partial manhattan spheres are employed to determine the number of gates within groups N_A, N_B, and N_C [123]–[125]. In order to consider the particular conditions that apply to those gates located close to the periphery of the circuit, which do not result in partial manhattan spheres, the gates are separated into two categories. The first category is the "starting gates," which denotes those gates that can form a hemisphere with radius l. Typically, these gates are located close to the center of the circuit and the "nonstarting gates," which denote those gates that cannot form partial spheres of radius l, are roughly located close to the periphery of the circuit [126]. An example of starting and nonstarting gates for a single plane is shown in Figure 4-3. Thus, gate P can be a starting gate forming a partial circle with radius $l = 2$, depicted by the solid line. In addition, gate Q cannot be a starting gate for gate pairs at a distance $l = 2$. Alternatively, gate Q can be a starting gate for gate pairs at a distance $l = 1$, as indicated by the dashed curve. Gate S is a nonstarting gate as a circle cannot be formed with a gate located below this gate for any distance l. The same definition applies to the case of multiple physical planes where the circles are substituted by spheres.

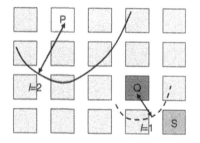

■ **FIGURE 4-3** Example of starting and nonstarting gates. Gates P and Q can be starting gates, while S is a nonstarting gate [126].

As in a 2-D wirelength model, a circuit element consists of only one gate. The gates included in groups N_A, N_B, and N_C, for a 3-D circuit, are [123]

$$N_A = 1, \tag{4-12}$$

$$N_B(l) = \sum_{i-1}^{l-1} N_c(i), \tag{4-13}$$

$$N_C(l) = \frac{M_{3-D}(l)}{N_{start}(l)}, \tag{4-14}$$

where $M_{3-D}(l)$ is the number of gate pairs at a distance l in a 3-D circuit,

$$M_{3-D}(l) = M_{2-D}(l) + 2\sum_{i=1}^{n-1} M_{2-D}(l - id_v). \tag{4-15}$$

The first term in (4-15) considers the intraplane gate pairs, while the second term includes the interplane gate pairs formed in the remaining n-1 planes. The necessary expressions to determine the number of starting gates $N_{start}(l)$ are presented in Appendix A.

Evaluating the starting and nonstarting gate pairs requires a significant number of summations. Alternatively, this differentiation between the different types of gates can be removed, resulting in a simpler approach to produce an interconnect length distribution for 3-D circuits as discussed in [127]–[129]. In this approach, an infinite sea of gates is considered, and the physical constraint that gates located at the periphery of the circuit cannot form a partial manhattan circle is relaxed. The average number of gates that belong to groups N_A, N_B, and N_C for a 3-D circuit is

$$N_A = 1. \tag{4-16}$$

$$
\begin{aligned}
N_B(l) \approx (l(l-1) &+ 2(l-d_v)(l-d_v-1) + l(l-1) \\
&+ 2(l-2d_v)(l-2d_v-1) + \ldots + l(l-1) + \ldots \\
&+ 2(l-(n-1)d_v)(1-(n-1)d_v-1) + \ldots + l(l-1))/n,
\end{aligned} \tag{4-17}
$$

$$
\begin{aligned}
N_C(l) = (2l &+ 4(l-d_v) + 2l + 4(l-2d_v)) + \ldots + 2l + \ldots \\
&+ 4(l-(n-1)d_v) + \ldots + 2l)/n,
\end{aligned} \tag{4-18}
$$

where n is the number of planes within the 3-D system and d_v is the interplane distance between two adjacent planes. The number of gate pairs that are l gate pitches apart, $M_{3\text{-}D}(l)$, is the sum of the gate pairs within a plane $M_{intra}(l)$ and the gate pairs in different physical planes $M_{inter}(l)$. By removing the distinction between starting and nonstarting gates, the boundary conditions of the problem are altered. Consequently, a normalization factor Γ' is required to ensure that the total number of interconnects within a circuit implemented in both two and three dimensions is maintained constant.

A similar distribution for interconnections in 3-D circuits can be determined by changing the number of gates that correspond to a circuit element, simplifying the enumeration process. Thus, for a 3-D circuit consisting of n planes, the expected interconnections between cell pairs (and not gate pairs) are enumerated [130], [131]. Due to this modification, the length distribution for the horizontal and vertical wires can be separately determined from the prediction model for 2-D circuits, as described in Section 4.1. Horizontal wires are considered only as intraplane interconnects in each plane of a 3-D circuit. Alternatively, the vertical wires include one or more interplane vias and can also include horizontal interconnect segments.

Since a cell is composed of n gates, the expected interconnections between cell pairs and not gate pairs are enumerated. In other words, the groups N_A, N_B, and N_C become, respectively,

$$N_A = n, \tag{4-19}$$

$$N_B(l) = \sum_{k=1}^{l-1} 2nk = nl(l-1), \tag{4-20}$$

$$N_C(l) = 2nl. \tag{4-21}$$

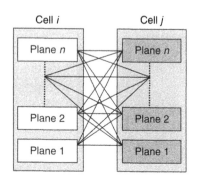

FIGURE 4-4 Possible vertical interconnections for two cells with each cell containing n gates [130].

In order to determine the distribution of the vertical wires, all possible connections between two cells are considered. The possible interconnects between a pair of cells is shown in Figure 4-4. Using (4-19)–(4-21) and by slightly modifying the expressions for the interconnect length distribution of the model presented in Section 4.1, the distribution of horizontal and vertical wires in 3-D circuits is obtained.

In all of the interconnect length distribution models for 3-D circuits presented in this section, an enumeration of the gate pairs is required. This process typically involves several finite series summations. Alternatively, a hierarchical approach can be adopted to generate the wirelength distribution within a 3-D circuit [132]. With this approach, the 3-D system is partitioned into K levels of hierarchy in descending order. The entire system is therefore partitioned into eight subcircuits at level K-1, while the lowest level (level 0) consists of single gates. The interconnect distribution of all of the interconnect lengths smaller than the maximum length (*i.e.*, $l \leq l_{max}$) is determined by the following relationship,

$$I(l) = S(l)q(l), \tag{4-22}$$

where $S(l)$ is the structural distribution of the number of all possible gate pairs. The structural distribution is physically equivalent to the number of gate pairs $M_{2\text{-}D}(l)$, as applied to the model presented in Section 4.1. The process of calculating $S(l)$ proceeds, however, with the method of generating polynomials [120]. A generating polynomial of a length distribution is equivalent to the moment generating polynomial function of that distribution and can be described by

$$g(x) = \sum_{l=0}^{\infty} S(l)x^l, \tag{4-23}$$

where each term provides the number of gate pairs connected with an interconnect of length l. Generating polynomials can provide a more efficient representation of the length distribution, as more complex interconnections among circuit elements can be generated by combining simple interconnection structures and exploiting symmetries that can exist within different circuit interconnection topologies. Consequently, the polynomials evaluated at the *kth* level of hierarchy can be used to construct the polynomials of the $(k + 1)th$ level. An analysis of the technique of generating polynomials is found in [133], [134]. The occupation probability $q(l)$, which is the probability of a gate pair being connected with an interconnect of length l, is based on Rent's rule [115] and Donath's approach [135] and, for a 3-D circuit, is

$$q(l) = Cl^{3(p-2)}, \qquad\qquad (4\text{-}24)$$

where C is a normalization constant in proportion to that constant used in the 2-D prediction model, as discussed in the previous section.

Comparing (4-24) with (4-7), a considerable decrease in the number of interconnects is noted, as this number decays faster as a function of length (*i.e.*, a cubic relationship rather than a quadratic function of length). This faster decay illustrates the reduction in the number of long global interconnects in 3-D circuits. Exploiting this decrease in interconnect length, three-dimensional integration is expected to offer significant improvements in several circuit design characteristics, as discussed in the following section.

4.3 **PROJECTIONS FOR 3-D ICs**

A priori interconnect prediction models described in the previous section provide an important tool to estimate the behavior of 3-D systems and, consequently, assess the potential of this novel technology. The introduction of the third dimension considerably alters the interconnect distribution in ICs. The wirelength distribution of a four million gate circuit for different number of planes is illustrated in Figure 4-5 [123]. By increasing the number of planes, the length of the global interconnects decreases. In addition, the number of these global interconnects decreases, while the number of short local interconnects increases to ensure that the total number of interconnects is essentially conserved.

The third dimension can be used to improve various characteristics of an IC [136], [137]. For example, tradeoffs among the maximum clock frequency, the required number of metal layers, and the area of an IC can be explored to achieve different design objectives. Consider a multitier interconnect architecture where the metal lines in each tier have a different aspect ratio and pitch. In addition, each tier consists of orthogonally routed multiple metal layers.

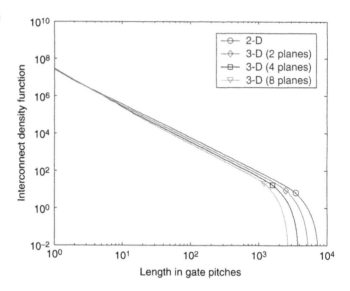

■ **FIGURE 4-5** Interconnect length distribution for a 2-D and 3-D IC [123].

As a smaller area per plane is feasible in a 3-D IC, the length of the corner-to-corner interconnects significantly decreases. Consequently, for a constant clock frequency, several global interconnects on the upper tiers can be transferred to the local tier with a smaller aspect ratio. The number of metal layers can therefore be reduced in 3-D circuits. Alternatively, if the number of metal layers is not reduced by transferring the global interconnects to the local tier interconnects, the clock frequency naturally increases as the longest distance for the clock signal to traverse is drastically reduced. For heterogeneous 3-D ICs, the clock frequency is shown to increase as $n^{1.5}$, where n is the number of planes within a 3-D IC [138]. Furthermore, the third dimension can be used to reduce the total area of an IC. Implementing a 2-D circuit in multiple planes results in a decrease in the length of the interconnects, creating a timing slack. If the clock frequency is maintained the same as that of a 2-D IC, the wiring pitch can be reduced, decreasing the total required interconnect area. The decrease in the interconnect pitch results in an increase in the resistive characteristics of the wires, which can consume the added timing slack provided by introducing the third dimension. Consider a 100 nm ASIC consisting of 16M gates; a 3-D IC with two planes produces a 3.9 × improvement in clock frequency, an 84% reduction in wire-limited area, and a 25% decrease in the number of metal layers for each plane within a 3-D circuit [124].

Since the introduction of the third dimension results, in general, in shorter interconnects, 3-D ICs consume less power as compared to 2-D

ICs. Using (4-12)–(4-15) and the expressions in [139] to generate the interconnect distribution for a 3-D system, a circuit composed of 16M gates is evaluated. In Figure 4-6, the variation of the gate pitch, total interconnect length, and power consumption are plotted versus the number of planes for two different values of the Rent's exponent. A value of $p = 0.5$ and $p = 0.6$ corresponds to a family of solid and dashed curves, respectively. Since higher values of p are related to systems with higher parallelism, a greater reduction in interconnect length can be achieved for these systems by increasing the number of planes, as indicated by the curves notated by a circle. Alternatively, highly serial systems, or equivalently, small values of the Rent's exponent permit higher device densities, since these circuits require fewer routing resources and, therefore, the gate pitch is significantly reduced with the number of planes. For greater values of p, the decrease in gate pitch is smaller, as illustrated by the dashed and solid curves notated by the square.

The interconnect power consumption depends on both the total interconnect length and the gate pitch. As both of these parameters are reduced with the number of planes, significant savings in power can be achieved with 3-D ICs. The sensitivity of the power consumption with respect to the Rent's exponent depends on the sensitivity of the gate pitch and the total interconnect length. Since these two parameters change in opposite directions with respect to the Rent's exponent, the sensitivity of the power consumption to this parameter is less significant (see the curves notated by the diamond in Figure 4-6). Overall,

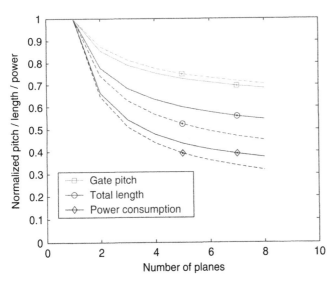

■ **FIGURE 4-6** Variation of gate pitch, total interconnect length, and interconnect power consumption with the number of planes [139].

an improvement in power consumption of 33%, 52%, and 62% is achieved for two, four, and eight planes, respectively [139].

This analysis only considers wire-limited circuits and ignores the power dissipated by the devices. A recent analysis for 3-D circuits designed with the MIT Lincoln Laboratories (MITLL) technology [140] also demonstrates the advantages of the third dimension. In this investigation of 3-D circuits [141], a Fast Fourier Transform (FFT) and an Open Reduced Instruction Set computer (RISC) Platform System-on-Chip (ORPSOC) [142] representing a low-power and a high-performance circuit example, respectively, are considered. The circuits are designed in a 180-nm SOI technology node. Predictions for other technology nodes are also extrapolated. The signal delay, power consumption, and silicon area for 2-D and 3-D versions of these circuits are listed in Tables 4-1 and 4-2, respectively. The clock skew and power consumption are reported in Table 4-3.

Table 4-1 Characteristics of 2-D circuits [141].

Technology Node	Circuit	Delay [ns]	Power [mW]	Area [mm²]
180 nm	FFT	25.81	1050	11.61
	ORPSOC	17.80	3298	18.66
90 nm	FFT	21.14	439	4.18
	ORPSOC	14.82	1944	6.76
45 nm	FFT	14.16	334	2.32
	ORPSOC	9.16	1215	3.26

Table 4-2 Characteristics of 3-D circuits [141].

Technology node	Circuit	Delay [ns]	Power [mW]	Area [mm²]
180 nm	FFT	21.49	952	4.01
	ORPSOC	14.78	2933	6.72
90 nm	FFT	19.54	394	1.48
	ORPSOC	10.02	1551	2.42
45 nm	FFT	13.01	256	0.82
	ORPSOC	8.33	1029	1.29

Table 4-3 Clock skew and power consumption [141].

Technology Node	Circuit	Skew [ps]		Power [mW]	
		2-D	3-D	2-D	3-D
180 nm	FFT	264.3	213.4	240.5	201.9
	ORPSOC	469.6	260.6	838.9	482.9
90 nm	FFT	167.0	88.8	61.0	44.6
	ORPSOC	317.6	213.0	694.2	419.1
45 nm	FFT	106.3	68.1	66.4	48.3
	ORPSOC	194.9	115.1	798.1	487.9

In addition to application-specific ICs (ASICs), general-purpose systems, such as microprocessors, are also expected to benefit considerably from vertical integration. As the speed and number of cores within a microprocessor system increase, the demand for larger and faster cache memories also grows. Implementing a larger cache memory inherently increases the latency of transferring data to the cores; the memory bandwidth, however, increases. In addition, cache misses become highly problematic. As the total area of a microprocessor system increases, the required time for transferring data from the processor to the main memory, due to a cache miss, also increases. Three-dimensional integration can significantly reduce the burden of a cache miss [143], [144]. Realizing a microprocessor system in multiple physical planes decreases the length of the data and address busses used for transferring data between the memory and the logic. In addition, by placing the memory on the upper planes, which consumes significantly lower power as compared to the power consumed by the cores, the power density can be confined within acceptable levels that will not exacerbate the overall cost of the system. The considerable improvements in speed and the power consumed by a 3-D microprocessor-memory system are discussed in Chapter 9.

4.4 **SUMMARY**

Interconnect length distribution models for 3-D circuits are analyzed in this chapter. Based on these models, a variety of opportunities that vertical integration offers to the IC design process are discussed. The major points of this chapter are as follows:

- Several *a priori* interconnect prediction models based on Rent's rule have been developed for 3-D ICs, unanimously demonstrating a significant decrease in the interconnect length for these circuits.

- Rent's rule correlates the number of terminals of a circuit with the number of modules that constitute the circuit and the average number of terminals in each module. The granularity of the module can vary from a single gate to an entire subcircuit.

- The interconnect distribution within a 3-D circuit can be determined by various methods, such as exhaustive enumeration and the method of generating polynomials.

- Vertical integration drastically alters the distribution of the interconnect lengths in a circuit, such that the number and length of the global interconnects decrease while the number of local interconnects increases.

- By introducing the third dimension, a variety of circuit characteristics can be improved. For example, the performance can be increased assuming a constant total area and number of metal layers. The number of required metal layers can be decreased for a fixed clock frequency and total area. Furthermore, the total area can be reduced while maintaining a fixed clock frequency.

- For heterogeneous 3-D ICs, the clock frequency increases as $n^{1.5}$, where n is the number of planes within a 3-D IC.

Chapter 5

Physical Design Techniques for 3-D ICs

A variety of recently developed or emerging fabrication processes for 3-D systems are reviewed in Chapters 2 and 3. An effective design flow, however, for 3-D ICs has yet to be developed. This predicament is due to the additional complexity and related issues that emerge from introducing the third dimension to the integrated circuit design process. Existing techniques for 3-D circuits focus primarily on the back end of the design process. Floorplanning, placement, and routing techniques for standard cell 3-D circuits are presented in, respectively, Sections 5.1, 5.2, and 5.3. These techniques constitute early research efforts to tackle these important problems rather than a complete design flow for 3-D circuits. Furthermore, many of these techniques incorporate thermal objectives or are exclusively dedicated to mitigate thermal problems in 3-D ICs. Although all of these physical design techniques have several common characteristics, due to the importance of thermal issues in three-dimensional integration, a discussion of thermal management techniques is deferred to Chapter 6. In addition to physical design techniques, a discussion on layout tools for 3-D circuits is presented in Section 5.4. A short summary of existing design methodologies for 3-D circuits is included in the last section of this chapter.

5.1 FLOORPLANNING TECHNIQUES

The predominant design objective for floorplanning a 2-D circuit has traditionally been to achieve the minimum area or, alternatively, the maximum packing density while interconnecting these blocks with minimum length wires. Most floorplanning algorithms can be classified as either slicing [145] or nonslicing [146], [147]. Floorplanning techniques belonging to both of these categories have been proposed for 3-D circuits [149]–[152]. An efficient floorplanning technique for

3-D circuits should adequately handle two important issues: representation of the third dimension and the related increase in the solution space.

Certain algorithms incorporate the 3-D nature of the circuits, such as a 3-D transition closure graph (TCG) [148], sequence triple [149], and a 3-D slicing tree [150] where the circuit blocks are notated by a set of 3-D modules that determine the volume of the 3-D system. Utilizing such a notation for the circuit blocks, an upper bound for 3-D slicing floorplans is determined [153]. In 3-D slicing floorplans, a plane successively bisects the volume of the 3-D system. The upper bound for 2- and 3-D slicing floorplans is illustrated in Figure 5-1. The coefficient r in Figure 5-1 is the shape flexibility ratio and denotes the maximum ratio of the dimensions of the modules (*i.e.*, max(*width/height, depth/height, width/depth*). In general, 3-D floorplans result in larger unused space as compared to 2-D slicing floorplans primarily due to the highly uneven volume of the 3-D modules. For high flexibility ratios, however, this gap is considerably reduced, and, in certain cases, the upper bound for 3-D floorplans becomes smaller than in 2-D floorplans.

Notating the location and dimensions of these modules typically requires a considerable amount of storage. A 3-D circuit, however, consists of a limited number of planes where the circuit blocks can be placed. Consequently, a 3-D system can be described as an array

■ FIGURE 5-1 Area and volume upper bounds for two- and three-dimensional slicing floorplans are depicted by the solid and dashed curve, respectively, for a different-shaped aspect ratio. V_{total} (A_{total}) and V_{max} (A_{max}) are the total and maximum volume (area), respectively, of a 3-D (2-D) system [153].

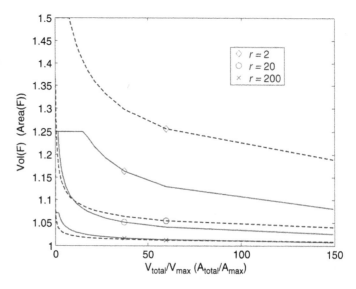

of two-dimensional planes where the circuit blocks are treated as rectangles that can be placed on any of the planes constituting the 3-D circuit [152]–[155]. The second challenge for 3-D floorplanning is to effectively explore the solution space, where a multistep approach can often be more efficient for floorplanning 3-D circuits than a flat approach, as discussed in the following subsection.

5.1.1 **Single-versus Multistep Floorplanning for 3-D ICs**

In 2-D circuits, a flat single-step approach is applied to generate a floorplan. Alternatively, in 3-D circuits, a multistep floorplanning technique is applied, where the blocks of a 3-D circuit are initially partitioned into planes of a stack [151]. A comparison between these two approaches is illustrated in Figure 5-2.

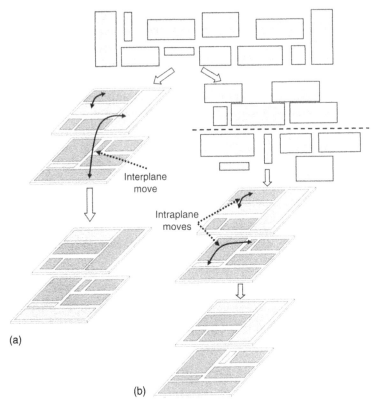

■ **FIGURE 5-2** Floorplanning strategies for 3-D ICs: (a) single-step approach and (b) multistep approach [156].

Interplane move

Intraplane moves

(a)

(b)

In single-step floorplanning algorithms, the floorplanning process proceeds by assigning the blocks to the planes of the stack followed by simultaneous intraplane and interplane block swapping, as depicted in Figure 5-2. Alternatively, a multistep approach does not allow interplane moves after the partitioning step. The reason for this constraint is that interplane moves among blocks result in a formidable increase in the solution space, directly affecting the computational time of a multistep floorplanning algorithm. Indeed, assuming N blocks of a 3-D system consisting of n planes, a flat floorplanning approach increases the number of candidate solutions by $N^{n-1}/(n\text{-}1)!$ times as compared to a 2-D circuit consisting of the same number of blocks. The solution space for floorplanning 2-D circuits based on the TCG technique [148], and 3-D circuits with a single-step and multistep approach are listed in Table 5-1. Consequently, a multiphase approach can be used to significantly reduce the number of candidate solutions.

The partitioning scheme adopted in the multistep approach plays a crucial role in determining the compactness of a particular floorplan, as interplane moves are not allowed when floorplanning the planes. Different partitions correspond to different subsets of the solution space, which may exclude the optimal solution(s). The objective function for partitioning should therefore be carefully selected. The partitioning can, for example, be based on minimizing the estimated total wirelength of the system [151]. Consequently, a partitioning problem based on minimizing the estimated total wirelength can be described as

$$\text{minimize} \sum_{net} EL_{net}, \tag{5-1}$$

$$\text{subject to } TN_{via} \leq V_{max}. \tag{5-2}$$

EL_{net} is the estimated interconnect length connecting two blocks of a 3-D circuit, which can contain both horizontal and vertical interconnect

Table 5-1 Solution space for 2-D and 3-D IC floorplanning [151].

		n-plane 3-D IC	
Characteristic	**2-D IC TCG**	**TCG 2-D array**	**Multi-step**
Solution space	$(N!)^2$	$N^{n-1}(N!)^2/(n-1)!$	$((N/n)!)^{2k}$
Ratio	1	$N^{n-1}/(n-1)!$	$1/\left(C_n^{n/k}\ldots C_{n/k}^{n/k}\right)^2$

segments. TN_{via} is the total number of interplane vias, and V_{max} is the maximum number of allowed interplane vias in a 3-D system. Optimizing (5-1) is a difficult task as linearity and convexity are not guaranteed, and, consequently, combinatorial techniques cannot be easily applied [151]. Hence, an algorithm based on simulated annealing (SA) can be utilized to minimize (5-1). The starting point for the SA engine is generated by randomly assigning the blocks to the planes of the system to balance the area of the individual planes. The SA process progresses by swapping blocks between two planes or changing the location of the blocks within one plane. The expected wirelength and number of vertical vias are reevaluated after each modification of the partition, where the algorithm progresses until the target solution is achieved at the desired low temperature of the SA algorithm.

As shown by (5-2), the allowed number of vias or, alternatively, the cut size, can affect the partition of the blocks. The effect of the cut size on the generated partition and the correlation between the cut size and the total wirelength can vary considerably in 3-D circuits [156]. A 3-D floorplanner based on [154] with a fixed cut size has been applied to the benchmark circuits included in the Microelectronic Center of North Carolina/Gigascale Research Center (MCNC/GSRC) benchmark suite [157]. Results of this evaluation indicate that partitioning is important for circuits where the interconnects among blocks exhibit a narrow wirelength distribution. Alternatively, circuits that include interconnects with a wide distribution of lengths are least affected by the partitioning step. In addition, results characterizing the relationship between the total wirelength and the allowed number of vertical interconnects across the planes of the stack show that the total interconnect length does not strongly depend on the cut size if the circuit consists of a small number of highly unevenly sized blocks. This behavior can be attributed to the significant computational effort required to optimize the area of the floorplan rather than the interconnect length. Alternatively, in circuits composed of uniformly sized blocks, an inverse relationship between the number of vertical vias and the interconnect length is demonstrated.

In the second phase of this technique, the floorplan of each plane of a 3-D circuit is generated. Note that the floorplan of each of the planes is simultaneously produced. The interconnects among those blocks that belong to different planes contribute to the total wirelength. Simulated annealing is also utilized in the majority of floorplanning algorithms, both 2-D and 3-D circuits. The circuit blocks are represented in three dimensions by the corner block list method [158]. Since the number of vertical vias is constrained by (5-2), the primary goal of the floorplanning step is to minimize the total wirelength and area. To achieve the desired low temperature of the SA algorithm, a candidate solution is perturbed by selecting a plane within the 3-D stack and applying one of the moves described in [158]. The multistep floorplanning technique is evaluated on the MCNC and GSRC benchmark suites [157] and compared to the TCG-based 2-D array and the combined bucket and 2-D array techniques (CBA) [155]. The results listed in Table 5-2 exhibit a small reduction, on the order of 3%, in the number of vertical vias and a significant reduction of approximately 14% in wirelength, while the total area increases by almost 4% for certain benchmark circuits.

5.1.2 Multi-Objective Floorplanning Techniques for 3-D ICs

The complexity of three-dimensional integration requires several dissimilar metrics for evaluating efficient floorplans for 3-D circuits beyond the traditional area and wirelength metrics. These metrics can consider, for example, the communication throughput among the circuit blocks or the number of interplane vias. A communication-based metric can be used in the design of microprocessors, resulting in floorplans with a higher number of instructions per cycle (IPC) [159].

In a 3-D system, blocks that communicate frequently are assigned to adjacent planes that decrease the required interconnect length of the interblock connections. The communication throughput is also increased while reducing the power consumed by the system. Alternatively, blocks with high switching activities should not overlap in the vertical direction to ensure that the temperature profile of the system remains within specified limits. Consequently, a 3-D floorplanner should carefully balance the communication throughput with the operating temperature.

In addition to the cut size or, equivalently, the number of interplane vias between planes, the processing technique used to bond the planes of a 3-D circuit also affects the partition step in a multistep floorplanning methodology. The various bonding mechanisms employed in a 3-D system contribute in different ways to the final floorplan. For example, front-to-front bonding provides a large number of interplane vias with extremely short lengths, improving the performance of those modules with a high switching activity. Furthermore, a block with a

Table 5-2 Multistep floorplanning results [151].

Benchmark	TCG-based 2-D Array			CBA			Multistep Floorplanning		
	Area	Wirelength	Vias	Area	Wirelength	Vias	Area	Wirelength	Vias
ami33	3.52E+05	23139	106	3.44E+05	23475	111	4.16E+05	21580	108
ami49	1.49E+07	453083	191	1.27E+07	465053	203	1.42E+07	420636	198
n100	53295	97066	704	51736	90143	752	54648	74176	733
n200	51714	198885	1487	50055	175866	1361	55944	142196	1358
n300	74712	232074	1613	75294	230175	1568	79278	213538	1534
Avg.	1.00	1.17	1.03	0.96	1.14	1.02	1.00	1.00	1.00

large area can be divided into two smaller blocks assigned to adjacent planes and employs front-to-front bonding in order to minimize the effect of the physical separation on the performance and power consumption of the block. Alternatively, interplane vias utilized in front-to-back bonding can adversely affect the performance of a 3-D system. The density of these vias is therefore low. Pairs of blocks that share a small number of interblock connections can be assigned to planes glued with front-to-back bonding.

Such a multi-objective floorplanning approach targeting microprocessor architectures is illustrated in Figure 5-3, where a variety of tools is utilized to characterize different parameters of the processor functional blocks. The CACTI [160] and GENESYS [161] tools provide an estimate of the speed, power, and area of the processor. The SimpleScalar simulator [162] combined with the Watch [163] framework records the information exchanged across the system to predict the power consumption of each benchmark circuit. Although a hierarchical approach is followed in this technique, the simulated annealing engine is replaced by a slicing algorithm based on recursive bipartitioning [164] to distribute the functional blocks of the processor onto the planes of the 3-D stack in order to decrease the computational time.

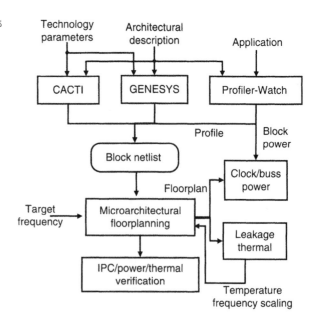

■ **FIGURE 5-3** Design flow of the microarchitectural floorplanning process for 3-D microprocessors [159].

Additional objectives include area and wirelength (A/W), area and performance (A/P), area and temperature (A/T), and area, performance, and temperature (A/P/T). Based on information gathered from evaluating MCNC/GSRC benchmarks, A/W achieves the minimum area as compared to the other objectives, decreasing by almost 40% the interconnect length over a 2-D floorplan of the microarchitecture. A/P increases the IPC by 18% over A/W, while simultaneously increasing the temperature by 19%. The more complex objective A/P/T generates a temperature close to A/W, while the IPC is increased by 14%. In general, the performance generated by the A/P/T objectives is bounded by the performance provided by the A/T and A/P objectives. In addition, A/P/T achieves higher performance as compared to A/W with similar temperatures [159].

5.2 PLACEMENT TECHNIQUES

Placement algorithms traditionally target minimizing the area of a circuit and the interconnect length among the cells, while reserving space for routing the interconnect. In vertical integration, a "placement dilemma" arises in deciding whether two circuit cells sharing a large number of interconnects can be more closely placed within the same plane or placed on adjacent physical planes, decreasing the interconnection length. Placing the circuit blocks on adjacent planes can often produce the shortest wirelength to connect these blocks. An exception is the case of small blocks within an SiP or SOP where the length of the interplane vias is greater than 100 μm. Since interplane vias consume silicon area, possibly increasing the length of some horizontal interconnects, an upper bound for this type of interconnect resource is necessary. Alternatively, sparse utilization of the interplane interconnects can result in insignificant savings in wirelength.

Several approaches can be adopted for placing circuit cells within a volume [165]–[169]. The circuit blocks can be treated, for example, as interconnected three-dimensional cells. This approach may not depict the discrete nature of a 3-D system as the circuit blocks can only be placed on a specific discrete number of planes; yet allows the formulation of a continuous, differentiable, and possibly convex objective function that can be optimally solved [166]. Since this approach does not consider the discrete number of planes available for circuit placement, a second step — referred to as a legalization step — is required to finalize the cell and interplane via placement of the planes within a 3-D circuit.

The length of a net connecting multiple cells can be described by the Euclidean distance among the cells connected by this net in three dimensions [166]. Alternatively, the distance of the terminals of a net can be adopted as the objective function to characterize the wirelength. To consider the effect of the interplane interconnects, a weighting factor can be used to increase the distance in the vertical direction, controlling the decision as to where to insert interplane vias. This weight essentially behaves as a control parameter that favors the placement of highly interconnected cells within the same or adjacent physical planes. In addition, algorithms that place interplane vias without overlaps and support a design rule compliant via placement are required [168]. To avoid overlap among interplane vias, a via placed within the bounding box of the net may not be possible. In this case, an increase in the wirelength is inevitable. A large weight is assigned to this additional interconnect segment to minimize the wire overhead. Although placing an interplane via within the bounding box of a net does not increase the wirelength, no existing placement techniques exist to determine the optimal location within this available region. This issue is discussed in Chapters 7 and 8, where the optimum via locations within available regions are determined.

5.2.1 Multi-Objective Placement for 3-D ICs

As with floorplanning, multi-objective placement techniques for 3-D circuits are of significant importance. Additional objectives that affect the cell placement and wirelength are simultaneously considered. Such objectives can include the temperature of the circuit and power supply noise (simultaneous switching noise (SSN)). The objective function optimized in this case is described by [170]

$$w_1 A^{total} + w_2 W^{total} + w_3 D^{total} + w_4 T^{total}, \tag{5-3}$$

where A^{total} is the total area of the 3-D system, W^{total} is the total wirelength, D^{total} is the required amount of decoupling capacitance to satisfy target noise margins, and T^{total} is the maximum temperature of the substrate. The w_i terms denote user-defined weights that control the importance of each objective during the placement process.

In order to manage these different objectives, additional information describing the circuit blocks is required including: *(i)* the current signature of each block, *(ii)* the number of metal layers dedicated for the power distribution network and the number and location of the power/ground pins, and *(iii)* the allowed voltage ripple on the power/ground lines due to simultaneous switching noise. Based on this information, the required amount of decoupling capacitance for each circuit can be determined.

This decoupling capacitance is distributed to the neighboring white spaces (*i.e.*, open areas not occupied by circuit cells). These spaces include those areas not only within the same plane but also on adjacent physical planes. To detect the white space within each plane, a vertical constrained graph is utilized. The upper boundary of the blocks at the *ith* level of the tree is compared to the lower boundary of the blocks at the $(i + 1)th$ level of the tree. An example of detecting white space is illustrated in Figure 5-4.

Although the white space can be extended to the adjacent planes in a 3-D IC, the decoupling capacitance allocated to these spaces may not be sufficient to suppress the estimated SSN in certain blocks. In these cases, the white space is expanded in the *x* and *y* directions to accommodate additional decoupling capacitance. The expansion procedure is depicted in Figure 5-5.

Several optimization techniques can be employed to efficiently determine the available white regions and allocate decoupling capacitance within these regions. Simulated annealing (SA) is an optimization algorithm typically used for this purpose, while different techniques for representing the circuit blocks can be adopted, such as the sequence pair technique [171]. To reach the SA freezing temperature, the solution generated at each iteration of the simulated annealing algorithm is perturbed by swapping operations between pairs of blocks. These perturbations include both intraplane and interplane swapping.

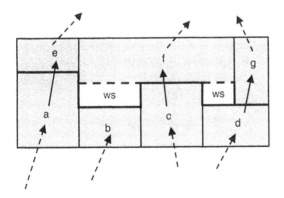

■ **FIGURE 5-4** White space detection is illustrated by the white regions [172].

■ **FIGURE 5-5** Block placement of an SOP: (a) an initial placement and (b) an increase in the total area in the x and y directions to extend the area of the white spaces [172].

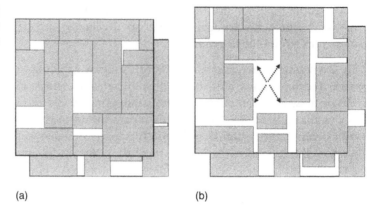

(a)   (b)

Although simulated annealing is a robust optimization technique, the effectiveness of such a multi-objective placement technique depends greatly on the accuracy of the physical model used to describe the various objectives in addition to the area and wirelength of the circuit interconnections. For example, the accuracy of the model describing the power distribution network of a 3-D system can either result in insufficient reduction in SSN or redundant decoupling capacitance, which increases the area, power, and total wirelength and therefore the cost of a circuit. Important traits of a power distribution network that should be considered for SSN suppression are the impedance characteristics of the paths to the current load (*i.e.*, the active devices), power supply pins, and decoupling capacitance. The accuracy can be further improved by including the parasitic series resistance and inductance of the decoupling capacitors and the inductive impedance of the interconnect paths [173]–[175].

Different objectives can also be added to the objective function in (5-3) without significantly modifying the solution procedure. For example, the same approach can be applied to minimize wire congestion in a 3-D SOP as well as area, wirelength, and SSN. The modified objective function in this case is described by [176]

$$w_1 A^{total} + w_2 W^{total} + w_3 D^{total} + w_4 C^{total}, \tag{5-4}$$

where C^{total} is the wire congestion. These multi-objective techniques have been applied to the GSRC [156] and GT benchmarks [177]. Some of these results are listed in Table 5-3, demonstrating the

Table 5-3 Placement for a four-plan SOP with diverse design objectives [170].

Circuits		Area/Wire Driven [mm², m, nF, °C]				Decap-driven [mm², m, nF, °C]				Multi-objective[mm², m, nF, °C]			
Name	Size	Area	Wire	Decap	Temp	Area	Wire	Decap	Temp	Area	Wire	Decap	Temp
n50	50	221	26.6	18.0	87.2	232	30.5	5.2	85.2	294	35.5	9.3	76.2
n100	100	315	66.6	78.2	86.5	343	73.1	69.2	81.7	410	77.0	77.9	78.5
n200	200	560	17.1	226.3	96.4	693	20.5	223.2	96.2	824	21.3	229.1	85.4
n300	300	846	28.6	393.8	100.1	843	28.6	393.8	100.1	844	28.6	393.8	100.1
gt100	100	191	13.2	60.8	71.0	207	16.8	42.5	70.9	264	18.6	55.2	59.2
gt300	300	238	19.6	342.5	93.2	248	22.3	334.9	99.5	256	22.3	343.9	85.3
gt400	400	270	28.1	493.1	114.0	268	32.5	482.0	111.6	282	34.6	492.6	91.1
gt500	500	316	30.3	645.3	99.7	321	35.4	632.4	98.0	321	34.8	635.8	95.8
Ratio		1.00	1.00	1.00	1.00	1.07	1.13	0.97	0.99	1.18	1.19	0.99	0.90

tradeoffs among the different design objectives [170], [172], [176]. The improvement in wire congestion is about 4%, independent of other design objective(s).

5.3 ROUTING TECHNIQUES

One of the first routing approaches for 3-D ICs demonstrated the complexity of the problem, as compared to the simpler single-device layer and multiple interconnect layer routing case [178]. Alternatively, recent investigations related to channel routing in 3-D ICs show that the problem is NP-hard [179]. Consequently, different heuristics have been considered to address routing in the third dimension [180], [183].

A useful approach for routing 3-D circuits is to convert the routing interplane interconnect problem into a 2-D channel routing task, as the 2-D channel routing problem has been efficiently solved [181], [182]. A number of methods can be applied to transform the problem of routing the interplane interconnects into a 2-D routing task, which requires utilizing some of the available routing resources for interplane routing. Interplane interconnect routing can be implemented in five major stages, including interplane channel definition, pseudo-terminal allocation, interplane channel creation (channel alignment), detailed routing, and final channel alignment [180]. Additional stages route the 2-D channels, for both the interplane and intraplane interconnects, and channel ordering to determine the wire routing order for the 2-D channels.

Each of these stages includes certain subtleties that should be separately considered. For example, a 3-D net should have two terminals in the interplane channel, and, therefore, inserting pseudo-terminals may be necessary for certain nets. In addition, aligning the channels may be necessary due to the different 2-D channel widths. Aligning the 2-D channels with adjacent planes of the 3-D system forms an overlapping region, which serves as an interplane routing channel. An example of such an alignment is shown in Figure 5-6. Since channel alignment can be necessary, at a later stage of the algorithm, the width of the 2-D channels is based on a detailed route of those channels without wires. Detailed channel routing with safe ordering follows all of the 2-D channels for both the intraplane and interplane interconnects, where a simulated annealing scheme is utilized. The technique is completed, if necessary, with a final channel alignment. Such an approach has been applied to randomly generated circuits and has produced satisfactory results [180].

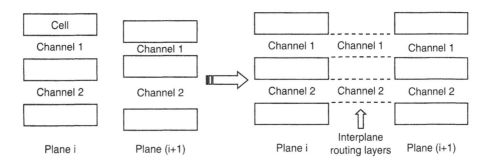

■ **FIGURE 5-6** Channel alignment procedure to create the interplane routing channels [180].

Although such a technique can offer a routing solution for standard cell and gate array circuits, alternative techniques that support different forms of vertical integration, for example, system-on-package (SOP), are required. In an SOP, the routing problem can be described as connecting the I/O terminals of the blocks located on the planes of the SOP through interconnect and pin layers. These layers, which are called routing intervals, are sandwiched between adjacent planes of an SOP. The structure of an SOP is illustrated in Figure 5-7, where each routing interval consists of pin redistribution layers and x-y routing layers between adjacent device layers. Communication among blocks located on nonadjacent planes is achieved through vias that penetrate the active device layers notated by the thick solid lines in Figure 5-7.

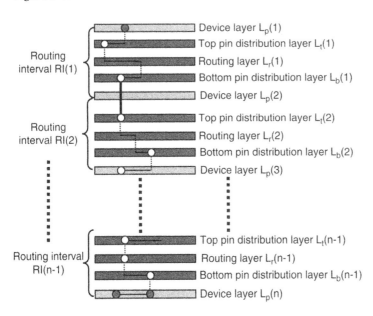

■ **FIGURE 5-7** An SOP consisting of n planes. The vertical dashed lines correspond to vias between the routing layers, and the thick vertical solid lines correspond to through silicon vias that penetrate the device layers [183].

For systems where the routing resources, such as the number of pin distribution layers, are limited, multi-objective routing is required to achieve a sufficiently small form factor. Other factors, such as integrating passive and active components, further enhance the demand for multi-objective routing approaches. A multi-objective approach can consider, for example, wirelength, crosstalk, congestion, and routing resources [183].

A function that accurately characterizes each of these objectives is necessary to produce an efficient route of each net n in an SOP. The wirelength can be described by the total manhattan distance in the x, y, and z-directions, where the z-direction describes the length of the interplane vias. The crosstalk produced from neighboring interconnects is

$$xt_n = \sum_{s \in NL, s \neq r} \frac{cl(r,s)}{|z(r) - z(s)|}, \tag{5-5}$$

where $cl(r,s)$ is the coupling length between two interconnects r and s, and $z(r)$ denotes the routing layer in which wire r is routed. The netlist that describes the connections among the nets of the blocks is denoted as NL. The delay metric used in the objective function for an interconnect r is the maximum delay of a sink of net r,

$$D^{\max} = \max\{d_r | r \in NL\}. \tag{5-6}$$

Finally, the total number of layers used to route an SOP consisting of n planes can be written as

$$L^{tot} = \sum_{1 \leq i \leq n} (|L_i(i)| + |L_r(i)| + |L_b(i)|). \tag{5-7}$$

For each routing interval i, $L_t(i)$, $L_r(i)$, and $L_b(i)$ denote the top pin distribution layer, routing layers, and bottom pin distribution layer, respectively. Combining (5-5)-(5-7), the global route of an SOP can proceed by minimizing the following objective function,

$$\alpha L^{tot} + \beta D^{\max} + \sum_{r \in NL} (\gamma xt_r + \delta wl_r + \epsilon via_r), \tag{5-8}$$

where via_r is the number of vias included in wire r, and the factors α, β, γ, and ϵ correspond to weights that characterize the significance of each objective during the routing procedure.

The various steps of a global routing algorithm optimizing the objective function described by (5-8) are illustrated in Figure 5-8. In order to distribute the pins to each circuit block, two different approaches can be followed: coarse (CPD) and detailed pin distribution (DPD).

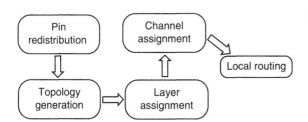

The difference between these two methods lies in the computational time. The complexity for CPD is $O(p \times u \times v)$, where p is the number of pins and $u \times v$ is the size of the grid on which the pins are distributed, while the complexity for DPD is $O(p^2 \ log \ p)$, exhibiting a quadratic dependence on the number of pins.

A topology (*i.e.*, Steiner tree) is generated for each of the interconnects within the routing interval to optimize the performance of the SOP. During the layer assignment process, the assignment of the routed wires is chosen to minimize the number of routing layers. The complexity related to the layer assignment is $O(N \ logN)$, where N is the total number of interconnects. The complexity of the channel assignment step is $O(|P|■|C|)$, where $|P|$ is the number of pins and $|C|$ is the number of channels. The algorithm terminates with a local route of the pins at the boundaries of the block and the pins within the routing intervals. Application of this technique to the GSRC and GT benchmark circuits exhibits an average improvement in routing resources of 35% with an average increase in wirelength of 14% as compared to routing that only minimizes wirelength. The maximum improvement can reach 54% with an increase in wirelength of 24% [183].

5.4 **LAYOUT TOOLS**

Beyond physical design techniques, sophisticated layout tools are a crucial component for the back-end design of three-dimensional circuits. A fundamental requirement of these tools is to effectively depict the third dimension and, particularly, the interplane interconnects. Different types of circuit cells for several 3-D technologies have been investigated in [184]. Layout algorithms for these cells have also been developed that demonstrate the benefits of 3-D integration. Other traditional features, such as impedance extraction, design rule checking, and electrical rule checking, are also necessary.

Visualizing the third dimension is a difficult task. The first attempt to develop tools to design 3-D circuits at different abstraction levels was introduced in 1984 [185], where symbolic illustrations of 3-D circuit cells at the technology, mask, transistor, and logic level are offered. A recent effort in developing an advanced toolset for 3-D ICs is demonstrated in [186] and [187] where the Magic layout tool has been extended to three dimensions (*i.e.*, 3-D Magic). To visualize the various planes of a 3-D circuit, each plane is illustrated on separate windows, while special markers are introduced to notate the interplane interconnects, as shown in Figure 5-9. In addition, impedance extraction is supported where the technology parameters are retrieved from predefined technology files. The 3-D Magic tool is also equipped with a reliability analysis design tool called ERNI 3-D [186]. This tool provides the capability to investigate certain reliability issues in 3-D circuits, such as electromigration, bonding strength, and interconnect joule heating. ERNI-3D is, however, limited to a two plane 3-D circuit structure.

A process design kit has recently been constructed by a research team from North Carolina State University [188] for designing 3-D circuits based on the MIT Lincoln Laboratories (MITLL) three-dimensional fabrication technology [140]. This kit is based on the commercial Cadence® Design Framework and offers several unique features for

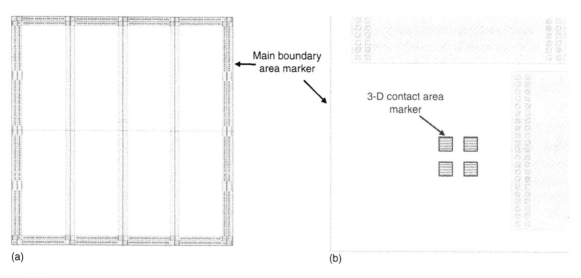

Main boundary area marker

3-D contact area marker

(a)   (b)

■ **FIGURE 5-9** Layout windows where different area markers are illustrated: (a) layout window for plane 1 and (b) layout window for plane 2 (windows are not of the same scale).

3-D circuits, such as visualizing circuits on the individual planes, highlighting features for interplane interconnects, and full 3-D design rule checking. The design rules for 3-D circuits have been largely extended to include aligning the interplane interconnects with the backside vias, which are an additional interconnect structure in the MITLL fabrication technology.

Although these tools consider important issues in developing a layout environment for 3-D circuits, there is significant room for improvement, as existing capabilities are rather rudimentary and limited to a specific 3-D technology and number of planes. In addition, a complete front- and back-end design flow for 3-D circuits does not yet exist. This situation is further complicated by the lack of a standardized fabrication technology for 3-D circuits. The complexity of 3-D integration poses significant obstacles in developing an efficient design flow. Finally, managing thermal effects, which are discussed in the next chapter, requires the thermal analysis process to be an inseparable element of the physical design process of 3-D integrated systems.

5.5 SUMMARY

The physical design techniques used during different stages of a developmental design flow for 3-D circuits are discussed, emphasizing the particular traits of 3-D ICs. The major points of this chapter are summarized as follows

- A variety of partitioning, floorplanning, and routing algorithms for 3-D circuits have been developed that consider the unique characteristics of 3-D circuits. In these algorithms, the third dimension is either fully incorporated or represented as an array of planes.

- The objective function within 3-D layout algorithms has been extended to include routing congestion, power supply noise, and decoupling capacitance allocation in addition to traditional objectives, such as wirelength and area.

- Simulated annealing is an essential optimization methodology within these algorithms as compared to greedy optimization techniques.

- Floorplanning algorithms usually consist of two basic steps: initial partitioning of the circuit blocks onto the planes of the stack and floorplan generation for each of the planes within the 3-D circuit.

- The partitioning step can significantly affect the quality of the solution as well as the computational time of the algorithms.

- In the second stage of the floorplanning algorithms where SA is applied, solution perturbations are realized by different intraplane moves among the blocks, such as swapping two blocks within a plane.

Chapter 6

Thermal Management Techniques

The primary advantage of three-dimensional integration, significantly greater packing density, is also the greatest threat to this emerging technology — aggressive thermal gradients among the planes within a 3-D IC. Thermal problems, however, are not unique to vertical integration. Due to scaling, elevated temperatures and hot spots within traditional 2-D circuits can greatly decrease the maximum achievable speed and affect the reliability of a circuit [189]. In addition, projected peak temperatures greatly deviate from International Roadmap for Semiconductors (ITRS) predictions of the maximum operating temperature in next-generation ICs [12]. Thermal awareness has therefore become another primary design issue in modern integrated circuits [190], [191].

In three-dimensional integration, low operating temperature is a prominent design objective, as thermal analysis of 3-D ICs indicates that escalated temperatures can be highly problematic [192]. In addition, peak temperatures within a 3-D system can exceed thermal limits of existing packaging technologies. Two key elements are required to establish a successful thermal management strategy: the thermal model, to characterize the thermal behavior of a circuit, and design techniques that alleviate thermal gradients among the physical planes of a 3-D stack while maintaining the operating temperature within acceptable levels. The primary requirements of a thermal model are high accuracy and low computational time, while thermal design techniques should produce high-quality circuits without incurring long computational design time. Although thermal models and techniques are separately discussed, the overall effectiveness of a thermal design methodology depends on both.

Consequently, this chapter emphasizes both the thermal models of 3-D ICs and the thermal design methodologies that have been recently developed. Models of different accuracy and computational speed are discussed in Section 6.1. Various techniques introduced at different steps of the IC design flow to alleviate projected thermal problems are also presented. An unorthodox yet potentially useful approach is applied here to discuss these techniques. The discussion is based on the objective used to decrease the temperature gradients within a 3-D system, rather than the specific design stage at which each technique is applied. Hence, those thermal strategies that aim to improve the thermal profile of a 3-D circuit without requiring any redundant interconnect resources for thermal management are presented in Section 6.2. Methodologies that are an integral part of a more aggressive thermal policy that utilize thermal vias, sacrificing other design objective(s), are outlined in Section 6.3. A synopsis of the primary points presented throughout this chapter is provided in Section 6.4.

6.1 THERMAL ANALYSIS OF 3-D ICs

A 3-D system consists of disparate materials with considerably different thermal properties, including semiconductor, metal, dielectric, and possibly polymer layers used for plane bonding. To describe the heat transfer process (only by conduction) within the volume of a system and determine the temperature at each point, T at a steady state requires the solution of

$$\nabla(k\nabla T) = -Q, \tag{6-1}$$

where k is the thermal conductivity and Q is the generated heat. In integrated circuits, heat originates from the transistors that behave as heat sources and also from self-heating of both the devices and the interconnects (joule heating), which can significantly elevate the circuit temperature [189], [192].

More specific techniques applied at various stages of the IC design flow, such as synthesis, floorplanning, and placement and routing, maintain the temperature of a circuit within specified limits or alleviate thermal gradients among the planes of the 3-D circuit. For each of the candidate solutions, (6-1) is solved for the entire volume of the system, requiring unacceptable computational time. To alleviate this issue so as to reduce the complexity of the modeling process, standard methods to analyze heat transfer, such as finite difference, finite element, and boundary element methods, have been adopted to evaluate the temperature of a 3-D circuit. Simpler analytic expressions have also been developed to characterize the temperature within a 3-D system. Each of these thermal models is described in the following subsections in ascending order of accuracy and complexity.

6.1.1 **Closed-Form Temperature Expressions**

Although thermal models based on analytic expressions tend to exhibit the lowest accuracy, these models can provide a coarse estimate of the thermal behavior of a circuit. This estimate may be of limited value at later stages of the design process where more accurate models are required; however, analytic models can be useful at the early stages of the design flow where physical information about the circuit does not exist. These first-order models can be used to determine various design aspects, such as packaging and cooling strategies and estimates of the overall system cost.

A 3-D system can be modeled as a cube consisting of multiple layers of silicon, aluminum, silicon dioxide, and polyimide, as shown in Figure 6-1. The devices on each plane are considered as isotropic heat sources and are each modeled as a thin layer on the top surface of the silicon layer. In addition, due to the short height of the 3-D stack, one-dimensional heat flow is assumed. Such an assumption vastly decreases the simulation time. Certain boundary conditions apply to this thermal model in order to validate the assumption of one-dimensional heat flow. Consequently, the lateral boundaries of a 3-D IC are considered to behave adiabatically (*i.e.*, no heat is exchanged with the ambient through the sidewalls), which is justified as the sidewalls of a package typically consist of insulating materials. The same assumption usually applies for the top surface of a 3-D circuit. Alternatively, the bottom side, which is typically connected to a heat sink, is treated as an isothermal surface. The insertion of these conditions limits the solution space, decreasing the required thermal profiling time. In addition, these boundary constraints capture the physical nature of the 3-D IC, preventing unreasonable temperatures.

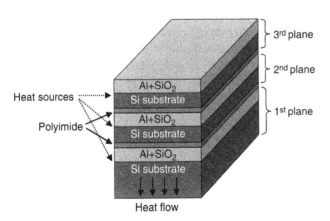

■ **FIGURE 6-1** Thermal model of a 3-D circuit where one-dimensional heat transfer is assumed [192].

Note that these boundary conditions apply to the thermal models for 3-D ICs discussed in this chapter, independent of the accuracy and computational time of these models.

Self-heating of MOSFET devices can also cause the temperature of a circuit to significantly rise. Certain devices can behave as hot spots, causing significant local heating. For a two-plane 3-D structure, an increase of 24.6°C is observed, due to the silicon dioxide and polyimide layers acting as thermal barriers for the flow of heat toward the heat sink [192]. Although the dielectric and glue layers behave as thermal barriers, the silicon substrate of the upper planes spreads the heat, reducing the self-heating of the MOSFETs. Simulation results indicate that by reducing the thickness of the silicon substrate from 3 μm to 1 μm in a two-plane 3-D IC, the temperature rises from 24.6°C to 48.9°C [192]. Thicker silicon substrates, however, decrease the packaging density and increase the length of the interplane interconnects. In addition, high aspect ratio vias can be a challenging fabrication task, as discussed in Chapter 3. If the silicon substrate is completely removed as in the case of 3-D SOI circuits, self-heating can increase to about 200°C, which can catastrophically affect the operation of the ICs. In this model, the interconnects are implicitly included by considering a specific aluminum density within a dielectric layer, as depicted in Figure 6-1.

To estimate the maximum rise in temperature on the upper planes of a 3-D circuit while considering the heat removal properties of the interconnects, a simple closed-form expression based on one-dimensional heat flow has been developed. The temperature increase ΔT in a 2-D circuit can be described by

$$\Delta T = R_{th} P / A, \tag{6-2}$$

where R_{th} is the thermal resistance between the ambient environment and the actual devices, P is the power consumption, and A is the circuit area. The power density P/A, also notated as Φ, increases in 3-D circuits due to the smaller footprint of the circuit, assuming the power consumption remains unchanged from a 2-D circuit to a 3-D circuit. The thermal resistance also changes, permitting (6-2) to be written as

$$\Delta T_j = \sum_{i=1}^{j} \left[R_i \left(\sum_{k=i}^{n} \frac{P_k}{A} \right) \right], \tag{6-3}$$

where P_k and R_k are the power consumption and thermal resistance of plane k, respectively.

Assuming the same power consumption and thermal resistance for all but the first of the planes, the increase in temperature is described by [193]

$$\Delta T_n = \left(\frac{P}{A}\right)\left[\frac{R}{2}n^2 + \left(R_{ps} - \frac{R}{2}\right)n\right]. \tag{6-4}$$

The thermal resistance of the first plane includes the thermal resistance of the package and the silicon substrate,

$$R_{ps} = {}^{t_{si1}}/_{k_{si}} + {}^{t_{pkg}}/_{K_{pkg}}, \tag{6-5}$$

where t_{sil} is the thickness of the silicon substrate of the first plane and k_{si} is the thermal conductivity of the silicon. The thermal resistance of the upper plane k is

$$R_k = {}^{t_{sik}}/_{k_{sik}} + {}^{t_{dielk}}/_{k_{dielk}} + {}^{t_{ifacek}}/_{k_{facek}}, \tag{6-6}$$

where t_{sik}, t_{dielk}, and t_{ifacek} are the thickness and k_{sik}, k_{dielk}, and k_{ifacek} are the thermal conductivity of the silicon substrate, dielectric layers, and bonding interface, respectively, for plane k. From (6-4)–(6-6), the increase in temperature on the topmost plane for various number of planes and power densities of a 3-D system is illustrated in Figure 6-2 for typical values of thicknesses and thermal conductivities of the substrates, dielectrics, and bonding materials. As shown in Figure 6-2, the temperature increase has a square dependence on the number of planes and a linear relationship with the power density.

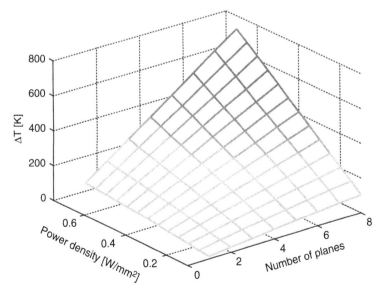

■ **FIGURE 6-2** Temperature increase in a 3-D circuit for different number of planes and power densities [192].

Note that the thermal resistance of the package contributes the greatest amount to the increase in temperature.

From Figure 6-2, the temperature within a 3-D circuit is exacerbated even with a small number of planes. The positive effect of the interconnects, particularly the interplane interconnects, on removing the heat and the interconnect joule heating has not been incorporated into the aforementioned expressions. The temperature increase on a specific plane N of a 3-D circuit considering the heat removal properties of the interconnect and the rise in temperature due to interconnect joule heating can be described by [194]

$$T_{si\_N} - T_{amb} = \sum_{i=1}^{N-1}\left[\sum_{r=1}^{N_i}\left(\sum_{s=1}^{N_i} j_{rms,ir}^2 \rho H_{ir} + \sum_{j=i+1}^{n}\Phi_j\right)\right] + \sum_{i=1}^{N}R_i\left(\sum_{k=i}^{n}\Phi_k\right),$$

(6-7)

where the first term describes the temperature increase from the interlayer dielectrics (ILDs), while the second term yields the rise in temperature caused by the package, the bonding materials, and the silicon substrate(s). The notations in (6-7) are defined in Table 6-1.

Expression (6-7) considers a 1-D model of the heat flow in a 3-D system similar to that shown in Figure 6-1. To analytically describe this heat flow, a thermal model, where the heat sources are represented as current sources and the thermal resistances as electrical resistors, is shown in Figure 6-3. The variation in temperature at different nodes of this thermal network resembles the node voltage in an electrical network where an Elmore delay-like model is used to determine the temperature at these nodes.

By including interplane vias and interconnect joule heating in the thermal model of a 3-D system, the thermal behavior of 3-D circuits can be more accurately modeled. The rise in temperature of a two-plane 3-D system is determined for two scenarios. In the former, interconnect joule heating and interplane vias are not considered, while in the latter, interconnect thermal effects are considered. A decrease of approximately 40° in the temperature of the bottom Si substrate is observed for the second scenario as compared to the first scenario. This result indicates the important role that interplane vias play on reducing the overall temperature of a 3-D system by decreasing the effective thermal resistance of the bonding and dielectric materials.

Although (6-7) includes the effect of the interconnect on the heat flow process, the various heat transfer paths that can exist within

Table 6-1 Definition of the symbols used in (6-7).

Notation	Definition
T_{amb}	Ambient temperature
n	Total number of planes
N_i	Number of metal layers in the ith plane
ir	r^{th} interconnect layer in the ith plane
t_{ILD}	Thickness of ILD
k_{ILD}	Thermal conductivity of ILD materials
s	Heat spreading factor
η	Via correction factor, $0 \leq \eta \leq 1$
j_{rms}	Root mean square value of current density for interconnects
ρ	Electrical resistivity of metal lines
H	Thickness of interconnects
Φ	Total power density on the mth plane, including the power consumption of the devices and interconnect joule heating
R_1	Total thermal resistance of package, heat sink, and Si substrate (bottom plane)
R_i $(i>1)$	Thermal resistance of the bonding material and the Si substrate for each plane

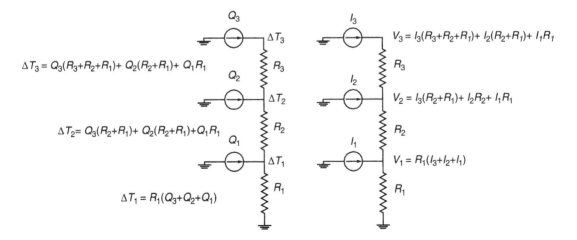

■ **FIGURE 6-3** An example of the duality of thermal and electrical systems.

interconnect structures have not been investigated. For example, assuming a one-dimensional heat flow, heat is only transferred through the vertical interconnects or, equivalently, the stacked interplane vias. Due to physical obstacles, such as circuit cells or routing congestion, a single vertical path may not be possible for certain interconnections. This situation is depicted in Figure 6-4 where different thermal paths are illustrated. As with current flow, heat flow also follows the path of the highest thermal conductivity. Consequently, interconnections consisting of horizontal segments in addition to interplane vias cause the heat flow to deviate from the vertical direction and spread laterally over a certain length, depending upon the length and thermal conductivity of each thermal path. By considering several thermal paths that can exist in a 3-D circuit, as shown in Figure 6-4, the effective thermal conductivity for the buried interconnect layer consisting of a dielectric and metal is [195]

$$k_{eff} = (1 - d_{vw})k_{ox} + d_{vw}k_{metal}, \qquad (6\text{-}8)$$

$$k_{eff} = k_{ox} + k_{metal.eff} = k_{ox} + \frac{t_{bi}}{A}\left[\frac{1}{R_1} + \frac{1}{R_2} + \frac{1}{R_3}\right]. \qquad (6\text{-}9)$$

where k_{ox}, k_{hm}, and k_{vm} are the thermal conductivity of the intralayer dielectric, intraplane interconnects, and interplane vias, respectively. t_{bi}, d_{vw}, and A are the thickness of the interconnect layer, the density of the interplane vias, and the area of the buried interconnect layer, respectively. The thermal resistance of the paths is notated by R_i, where these paths are considered in parallel, similar to electrical resistors connected in parallel. This duality implies that the presence of multiple thermal paths in a region results in a decrease in the total thermal resistance of that region.

As noted above, the self-heating of devices can affect the reliability of a circuit. Those devices that exhibit a high switching activity, such as clock drivers and buffers, can suffer considerably from local heating,

■ FIGURE 6-4 Different vertical heat transfer paths in a 3-D IC [195].

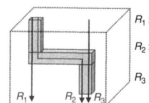

R_1 : Metal-dielectric

R_2 : Metal (horizontal and vertical)

R_3 : Dielectric-metal

resulting in degraded performance. By considering the various thermal paths that can exist within interconnect structures, the effect of a rise in temperature on these devices is investigated where the interplane interconnects are placed below the clock driver to reduce the self-heating process [195]. The increase in peak temperature as a function of the power density of the clock driver placed above different interconnect structures is illustrated in Figure 6-5. The existence of a thermal path with a horizontal metal segment exhibits inferior heat removal properties as compared to an exclusively vertical thermal path. In addition, the resulting increase in temperature in a 3-D IC is higher than bulk CMOS but not necessarily worse than SOI, as illustrated in Figure 6-5.

Another factor that can affect the thermal profile of a circuit is the physical adjacency of the devices. Thermal coupling among neighboring devices in a 3-D circuit is amplified, further increasing the temperature of the circuit [192], [195]. The temperature is shown to exponentially decrease with the gate pitch. This result means that the area consumed by certain circuit elements, such as a clock driver, should be greater in a 3-D circuit than in a 2-D circuit to guarantee reliable operation.

The assumption of one-dimensional heat flow permits a circuit to be modeled by a few serially-connected resistors. In addition, by

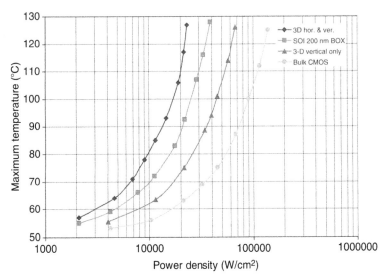

■ **FIGURE** 6-5 Maximum temperature vs. power density for 3-D ICs, SOI, and bulk CMOS [195]. The difference among the curves for the 3-D ICs is that the first curve (3-D horizontal and vertical) includes thermal paths with a horizontal interconnect segment while the second curve only includes the interplane vias (only 3-D vertical vias).

including the interconnect power and the various thermal paths, the accuracy of the thermal models of 3-D ICs is significantly improved. The major assumption and simultaneously drawback of the closed-form expressions describing the temperature of a 3-D circuit is that each physical plane is characterized by a single heat source. This assumption implies that all of the heat sources that can exist within a plane can be collapsed into a single heat source that is included in a thermal network similar to that depicted in Figure 6-3. Although this approach is sufficiently accurate at early steps in the design process, knowledge of the actual distribution of the power density and temperature within each physical plane is essential to thermal design methodologies in order to maintain thermally tolerant circuit operation. More accurate models required for such techniques are presented in the following subsection.

6.1.2 **Compact Thermal Models**

In the previous subsection, thermal models based on analytic expressions for the temperature of a 3-D circuit are discussed. In all of these models, the heat generated within each physical plane is represented by a single value. Consequently, the power density of a 3-D circuit is assumed to be a vector in the vertical direction (*i.e.*, z-direction). In addition, the thermal network is represented as a 1-D resistive network.

The temperature and heat within each plane of a 3-D system, however, can vary considerably, yielding temperature and power density vectors that depend on all three directions. Compact thermal models capture this critical information by representing the volume of a circuit with a set of nodes. The nodes are connected through resistors forming a 3-D resistive network. As an example of this model, consider a 3-D system modeled as a thermal resistive stack, as shown in Figure 6-6. The thermal network shown in Figure 6-6a is segmented into single pillars [196]–[198]. Each pillar is successively modeled by a 1-D thermal network including thermal resistors and heat sources, as shown in Figure 6-6c. The heat sources include all of the heat generated by each of the devices contained in each tile. The resistors embedded in the gray shaded triangles are those resistances related to the interplane vias. The absence of a via between two planes is incorporated by those via resistances to ensure that heat flow will not occur through those resistors. The voltage source at the bottom of the network models the isothermal surface between the heat sink and the bottom silicon substrate. Additional resistors, not

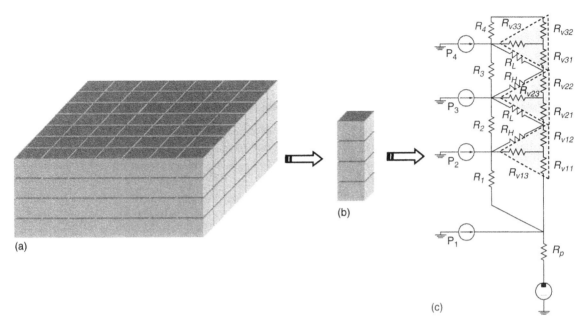

(a)

(b)

(c)

■ **FIGURE 6-6** Thermal model of a 3-D IC: (a) a 3-D tile stack, (b) one pillar of the stack, and (c) an equivalent thermal resistive network. R_1 and R_p correspond to the thermal resistance of the thick silicon substrate of the first plane and the thermal resistance of the package, respectively [197].

shown here, are used to incorporate the flow of heat among neighboring pillars. These resistors are treated as constant because the heat flow in the lateral directions is typically significantly smaller as compared to the vertical flow of heat.

Comparing the compact model of a single stack with the simpler 1-D model used to produce the closed-form solution presented in subsection 6.1.1, note that several similarities exist. Both of these models consist of resistors and heat sources modeled as current sources. Note also that although the voltage source included in Figure 6-6c, which considers the heat sink, does not appear in Figure 6-3, this element of the model is implicitly included in (6-7). This behavior occurs since the closed-form expression provides the temperature rise in plane N of the 3-D system (*i.e.*, $\Delta T = T_{si\_N} - T_{amb}$) rather than the absolute temperature generated by the compact model.

Compact thermal models provide more accurate temperature estimates as compared to analytic expressions, but are less accurate than the models discussed in the following subsection. Other issues specific to these compact models are that these models are independent of the boundary conditions of the target system [199]. In addition, these models should be scalable to be widely applicable. A possible solution to this requirement is appropriately scaling the resistances of the network for each 3-D technology when calibrating the model [197].

6.1.3 Mesh-Based Thermal Models

This type of model most accurately represents the thermal profile of a 3-D system. The primary advantage of mesh-based thermal models is that these models can be applied to any complex geometry and do not depend on the boundary conditions of the problem. A 3-D circuit is decomposed into a 3-D structure consisting of finite hexahedral elements (*i.e.*, parallelepipeds), as illustrated in Figure 6-7. The mesh can also be nonuniform for regions with complex geometries or nonuniform power densities.

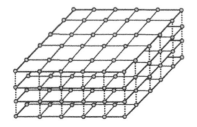

FIGURE 6-7 A four plane 3-D circuit discretized into parallelepipeds.

The temperature at the vertices of each parallelepiped, called a node, is evaluated from (6-1). By considering that the thermal conductivity k is independent of the temperature, (6-1) can be written as a linear function of temperature,

$$k\nabla^2 T = -Q. \tag{6-10}$$

To solve this differential equation, several numerical methods, such as the finite element method (FEM), the finite difference method, and the boundary element method, have been adopted [190], [200], [201]. The temperature at any other point within a 3-D mesh is determined from different interpolation schemes. For example, the temperature at a point within the parallelepiped shown in Figure 6-7 is determined from

$$T(x, y, z) = \mathbf{Nt} = \sum_{i=1}^{8} N_i t_i, \tag{6-11}$$

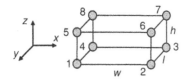

FIGURE 6-8 Fundamental parallelepiped used to model thermal effects in a 3-D IC based on FEM.

where t_i is the temperature and N_i is the shape function of node i. The shape function depends on two factors; the coordinates at the center of the element (x_c, y_c, z_c) and the dimensions of the element, as noted in Figure 6-8 and described by

$$N_i = \left(\frac{1}{2} - \frac{2(x_i - x_c)}{w^2}(x - x_c)\right)\left(\frac{1}{2} + \frac{2(y_i - y_c)}{l^2}(y - y_c)\right)\left(\frac{1}{2} + \frac{2(z_i - z_c)}{h^2}(z - z_c)\right). \tag{6-12}$$

As compared to compact thermal models, which include a network of thermal resistances characterizing the target 3-D system, these models provide greater accuracy in evaluating the temperature of the circuit since the model does not include any fixed resistances. Furthermore, the temperature at any point of the system can be determined by interpolating a detailed thermal profile of a 3-D circuit. If these models are embedded in an iterative thermal management technique, however, the thermal profile of a 3-D circuit can be obtained at every iteration, although this process is computationally expensive. Finally, due to the generality of these models, the thermal properties of the disparate materials comprising an integrated circuit are inherently considered. In the following sections, techniques are discussed that utilize these models to mitigate the effects of high thermal conditions in 3-D ICs.

6.2 THERMAL MANAGEMENT TECHNIQUES WITHOUT THERMAL VIAS

As with 2-D ICs, an increase in temperature can severely affect the performance of a 3-D circuit, since the interconnect resistance increases and the output current of the devices is reduced. Furthermore, as the leakage power is exponentially dependent on the temperature, an increase in temperature triggers a positive feedback mechanism, which can lead to thermal runaway. In addition, higher operating temperatures degrade the reliability, which can decrease the lifetime of a circuit. The cost of a circuit can also increase as more expensive packaging is required to handle high temperatures.

These issues are expected to be more pronounced in 3-D circuits due to the higher power density. This increase is also enhanced by the greater distance of the heat sources (*i.e.*, devices) on the upper planes, far from the heat sink. Finally, the reduced area of a 3-D system results in a smaller area for the heat sink, reducing the heat transferred to the ambient environment [202].

Several floorplanning, placement, and routing techniques for 3-D ICs and SOP have been developed that consider the high temperatures and thermal gradients developed throughout the planes of these systems beyond the traditional objectives, such as area and wirelength minimization. In Section 6.2.1., thermal-driven floorplanning techniques are discussed, while thermal-driven placement techniques are analyzed in Section 6.2.2.

6.2.1 Thermal-Driven Floorplanning

Traditional floorplanning techniques for 2-D circuits typically target the optimization of an objective function that includes the total area of the circuit and the total wirelength of the interconnections among the circuit blocks. For 3-D circuits, a further requirement for floorplanning is minimizing the number of interplane vias to decrease the fabrication cost and silicon area. Consequently, an objective function for 3-D circuit floorplanning can be written as

$$\cos t = c_1 \; wl + c_2 \; area + c_3 \; iv, \qquad (6\text{-}13)$$

where c_1, c_2, and c_3, are weight factors and wl, area, and iv are the normalized wirelength, area, and number of interplane vias [203].

A thermal-driven floorplanning technique would extend this function to include the thermal objective,

$$\cos t = c_1 \; wl + c_2 \; area + c_3 \; iv + c_4 g(T), \qquad (6\text{-}14)$$

where the last term is a cost function of the temperature. An example of a function where the cost is a ramp function of the temperature is schematically shown in Figure 6-9 [203]. Note that the cost function does not intersect the abscissa but rather reaches a plateau. Consequently, this objective function does not minimize the temperature of the circuit but rather constrains the temperature within specified levels. Indeed, minimizing the circuit temperature may not be an effective objective leading to prohibitively long computational times or to excessive increases in other design objectives.

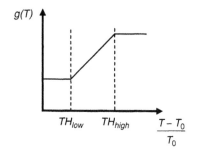

■ FIGURE 6-9 Temperature cost function [203].

Another important issue for physical design techniques in 3-D circuits is the representation of the third dimension, as also discussed in Chapter 5. A nonefficient representation scheme can considerably increase the storage requirements and, consequently, slow down the optimization process, regardless of the objective function.

A low overhead representation scheme is realized by representing the blocks within a 3-D system with a combination of 2-D matrices that correspond to the planes of the system and a bucket structure that contains the connectivity information for the blocks located on different planes (a combined bucket and 2-D array (CBA)) [203]. A transitive closure graph is used to represent the intraplane connections of the circuit blocks. The bucket structure can be envisioned as a group of buckets imposed on a 3-D stack. The indices of the blocks that intersect with a bucket are included in this bucket, regardless of the plane on which a block is located. A 2 × 2 bucket structure applied to a two-plane 3-D IC is shown in Figure 6-10, where the index of the bucket is also depicted. To explain the bucket index notation, consider the lower left tile of the bucket structure shown in Figure 6-10c (*i.e.,* b21). The indices of the blocks that intersect with this tile on the second plane are d and e, and the indices of the blocks from the first plane are l and k. Consequently, b21 includes d, e, l, and k.

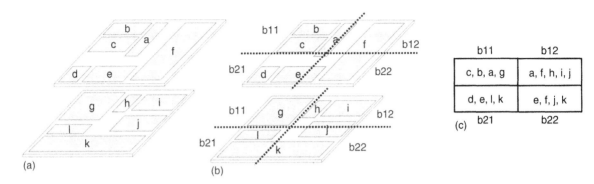

■ **FIGURE 6-10** A bucket structure example for a two-plane circuit consisting of 12 blocks: (a) a two plane 3-D IC, (b) a 2 × 2 bucket structure imposed on a 3-D IC, and (c) the resulting bucket indices [203].

Simulated annealing (SA) is a well-known method to produce an effective solution of an objective function, as in (6-14) for floorplanning 3-D circuits. The SA scheme converges to the desired freezing temperature through several solution perturbations. These perturbations include one of the following operations, some of which are unique to 3-D ICs:

- block rotation
- intraplane block swapping
- intraplane reversing of the position of two blocks
- movement of a block within a plane
- interplane swapping of two blocks
- z-neighbor swap
- z-neighbor move

The last three operations are unique to 3-D ICs, while the z-neighbor swap can be considered as a special case of interplane swapping of two blocks. Thus, two blocks located on adjacent planes are only swapped if the relative distance between these two blocks is small. In addition, the z-neighbor move considers the move of a block to another plane of the 3-D system without significantly altering the x-y coordinates. Examples of these two operations are illustrated in Figure 6-11.

In an exhaustive approach, each of the aforementioned block perturbations requires a thermal profile of a 3-D circuit. Such a tedious approach greatly increases the computational time of the algorithm. A thermal profile, therefore, is invoked after specific operations or after a specified number of iterations. The reasoning behind this practice is that certain operations, such as the move of two intraplane

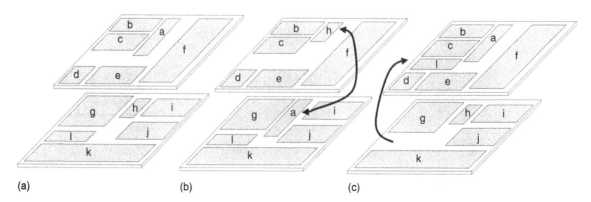

(a)   (b)   (c)

■ **FIGURE 6-11** Interplane moves: (a) an initial placement, (b) a z-neighbor swap between blocks a and h, and (c) a z-neighbor move for block I from the first plane to the second plane.

blocks or the rotation of a block, are not likely to significantly affect the temperature of a system, whereas other operations, such as a z-neighbor swap or a z-neighbor move, is expected to considerably affect the temperature of some blocks.

As discussed in Section 6.1, the efficiency of the thermal techniques depends significantly on the time required for evaluating the temperature within the volume of a 3-D circuit. Thermal models with different accuracy and computational time have been applied to MCNC benchmarks in conjunction with this floorplanning technique. Related results are reported in Table 6-2 where both the detailed and compact thermal modeling approach is considered [203]. A significant tradeoff

Table 6-2 Temperature decrease through thermal-driven floorplanning [203].

Circuit	CBA w/o Thermal Objective		CBA-T		CBA-T-fast	
	T [°C]	Runtime [sec]	T [°C]	Runtime [sec]	T [°C]	Runtime [sec]
ami33	471	23	160	466	204	56
ami49	259	86	151	521	196	144
n100	391	313	158	4322	222	446
n200	323	1994	156	6843	242	4474
n300	373	3480	167	17484	208	4953
Avg	1	1	0.44	9.71	0.6	1.82

between the runtime and the decrease in temperature exists between these thermal models. With thermal-driven floorplanning, where a grid of resistances is utilized to thermally model a 3-D circuit, a 56% reduction in temperature is achieved. The computational time, however, is increased by approximately an order of magnitude as compared to conventional floorplanning algorithms. Alternatively, if a closed-form expression is used for the thermal model of a 3-D circuit, the decrease in temperature is only 40%. The computational time, however, is approximately doubled in this case. Other design characteristics, such as the area and wirelength, do not change significantly between the two models.

In an effort to decrease the computational time, floorplanning can be performed in two separate phases. In the first step, the circuit blocks are assigned to the planes of the 3-D system to minimize area and wirelength. This phase, however, can result in highly uneven areas among the planes. A second step that limits these imbalances is therefore necessary. The objective function to accomplish this balancing process can be described as

$$\cos t = c_1 \cdot wl + c_2 \cdot area + c_3 \cdot dev(F) + c_4 \cdot P + c_5 \cdot TOP, \qquad (6\text{-}15)$$

where c_1, c_2, c_3, c_4, and c_5 notate some weighting factors. Beyond the first two terms that include the area and wirelength of the circuit, the remaining terms consider other possible design objectives for 3-D circuits. The third term is intended to minimize the imbalance that can exist among the dimensions of the planes within the stack, based on the deviation dimension approach described in [154]. Planes with particularly different areas or greatly uneven dimensions can result in a significant portion of unoccupied silicon area on each plane.

The last two terms in (6-15) consider the overall power density within a 3-D stack. The fourth term considers the power density of the blocks within the plane as in a 2-D circuit. The cost function characterizing the power density is based on a similarly shaped function as that used for the temperature depicted in Figure 6-9. Thermal coupling among the blocks on different planes is considered by the last term and can be written as

$$TOP = \sum \left(\sum_i P_i + P_{ij} \right), \qquad (6\text{-}16)$$

where P_i is the power density of block i and P_{ij} is the power density due to overlapping block i with block j from a different plane. The summation operand sums the contributions from the blocks located on all of the other planes other than the plane containing block j. If a simplified thermal model is adopted, an analytic expression as in (6-16) can be utilized to capture thermal coupling among the blocks, thereby compensating for the loss of accuracy originating from the thermal model.

This two step floorplanning technique has been applied to several Alpha microprocessors [204]. Results indicate a 6% average improvement in the maximum temperature as compared to 3-D floorplanning without a thermal objective. In addition, comparing a 2-D floorplan with a 3-D floorplan, an improvement in area and wirelength of 32% and 50%, respectively, is achieved [205]. The peak temperature, however, increases by 18%, demonstrating the importance of thermal issues in 3-D ICs.

The reduction in temperature is smaller as compared with the one-step floorplanning approach. With a two-phase approach, the solution space is significantly curtailed, resulting in a decrease in the computational time. The interdependence, however, of the intraplane and interplane allocation of the circuit blocks is not captured, which can yield inferior solutions as compared to one-step floorplanning techniques.

Although simulated annealing is the dominant optimization scheme used in most floorplanning and placement techniques for 3-D ICs [170], [183], [203], genetic algorithms have also been utilized to floorplan 3-D circuits. As an example, consider the thermal-aware mapping of 3-D systems that incorporate network-on-chip (NoC) architectures [206]. The merging of three-dimensional integration with NoC is expected to further enhance the performance of interconnect limited ICs (the opportunities that emerge from combining these two design paradigms are discussed in greater detail in Chapter 9).

Consider the 3-D NoC shown in Figure 6-12. The objective is to assign the various tasks of a specific application to the processing elements (PEs) of each plane to ensure the temperature of the system and/or communication volume among the PEs is minimized. The function that combines these objectives is used to characterize the fitness of the candidate chromosomes (*i.e.*, candidate mappings), which can be described by

$$S = \frac{1}{a + \log(\text{max\_temp})} + \frac{1}{\log(\text{comm\_cost})}. \qquad (6\text{-}17)$$

As with traditional genetic algorithms, an initial population is generated [207]. Crossover and mutation operations generate chromosomes,

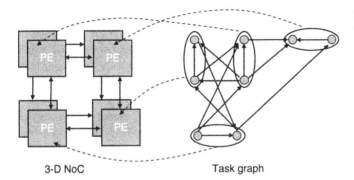

■ **FIGURE 6-12** Mapping of a task graph onto physical PEs within a 3-D NoC [206].

3-D NoC   Task graph

which survive to the next generation according to the relative chromosomal fitness. The floorplan with the highest fitness is selected after a number of iterations or if the fitness cannot be further improved.

In all of the techniques presented in this section, the heat is conveyed from the upper planes to the bottom plane primarily through the power and signal lines and the thinned silicon substrates. No additional means other than redistributing the major heat sources (*i.e.*, the devices) throughout the 3-D stack is utilized to decrease the overall temperature of the planes to alleviate significant thermal gradients. As previously discussed in Section 6.1, the interplane interconnects can carry a significant amount of heat toward the heat sink, reducing the temperature and the thermal gradients within a 3-D IC. Alternatively, the objective function in (6-14) aims at minimizing or at least limiting the number of interplane vias. These vias, however, accelerate the flow of heat to the ambient, in addition to connecting circuits located on different physical planes of the stack. Consequently, such methods ignore the benefit of the interplane vias.

6.2.2 **Thermal-Driven Placement**

Placement techniques can also be enhanced by the addition of a thermal objective. As discussed in Section 6.1, finite difference and finite element methods can be utilized to solve (6-1). With the finite difference approximation, (6-1) can be written as $R{\cdot}P = T$, where R is the thermal resistance matrix. The elements of the thermal matrix describe the thermal resistance between two nodes in a 3-D mesh, while the temperature T and power vector P describe the temperature and power dissipation, respectively, of each node.

Any modification in the placement of the cells causes all of these matrices to change. In order to determine the resulting change in temperature, the thermal resistance is updated and multiplied with the power density vector. This approach, however, leads to long computational times. Simpler and less accurate approaches should therefore be utilized. For each modification of the block placement, the change in the power vector ΔP is scaled by R, and the change in the temperature vector is evaluated. A new temperature vector is obtained after the latest move of the blocks within a 3-D system. Some results of a thermally driven placement are listed in Table 6-3, where the results from a placement based on traditional area and wirelength objectives are also provided for comparison [170].

The force-directed method is another technique used for cell placement [208], where repulsive or attractive forces are placed on the cells as if these cells are connected through a system of springs. The force-directed method is extended to incorporate the thermal objective during the placement process [209]. In this approach, repulsive forces are applied to those blocks that exhibit high temperatures (*i.e.*, "hot blocks") to ensure that the high-temperature blocks are placed at a greater distance from each other. To obtain the temperature of a 3-D circuit, the finite element method is utilized. The 3-D stack is modeled by a mesh consisting of unit parallelepipeds, as discussed in subsection 6.1.3, where the interpolation functions described by (6-11) and (6-12) are used to evaluate the temperature of the elements. The

Table 6-3 Thermal-driven floorplanning for four plane 3-D ICs [170].

Circuit	Area/Wire Driven [mm²,m,nF,°C]				Thermal Driven [mm²,m,nF,°C]			
	Area	Wire	Decap	Temp	Area	Wire	Decap	Temp
n50	221	26.6	18.0	87.2	377	84.1	29.7	68.9
n100	315	66.6	78.2	86.5	493	24.5	93.6	69.8
n200	560	17.1	226.3	96.4	1077	38.8	243.6	76.2
gt100	846	28.6	393.8	100.1	1310	20.4	405.3	86.6
gt300	191	13.2	60.8	71.0	474	28.0	92.7	52.3
gt400	238	19.6	342.5	93.2	528	37.0	392.1	72.1
gt500	270	28.1	493.1	114.0	362	38.5	512.0	89.2
gt600	316	30.2	645.3	99.7	541	76.5	684.4	80.3
Ratio	1.00	1.00	1.00	1.00	1.75	1.51	1.08	0.80

thermal gradient of a point or node of an element can be determined by differentiating (6-11),

$$\mathbf{g} = \left[\frac{\partial T}{\partial x} \ \frac{\partial T}{\partial y} \ \frac{\partial T}{\partial z}\right]^{\mathrm{T}}. \qquad (6\text{-}18)$$

Exploiting the thermal electric duality, the modified nodal technique in circuit analysis [210], where each resistor contributes to the admittance matrix (*i.e.*, matrix stamps), can also be utilized to construct element stiffness matrices for each element. These elemental matrices are combined into a global stiffness matrix for the entire system, notated as $\mathbf{K}_{\mathrm{global}}$. This matrix is included in a system of equations to determine the temperature of the nodes that characterizes the entire 3-D circuit. The resulting expression is

$$\mathbf{K}_{\mathrm{global}} \, \mathbf{T} = \mathbf{P}, \qquad (6\text{-}19)$$

where $\mathbf{P}$ is the power consumption vector of the grid nodes. To determine this vector, the power dissipated by each element of the grid is distributed to the closest nodes. From solving (6-19), the temperature of each node can be determined at every iteration of the force-directed algorithm.

The force-directed approach utilizes a function that is minimized through the application of repulsive forces on the hot elements. This function is described by a system of equations where the cost of a connection between two nodes as defined in [208] is used to produce another global stiffness matrix $\mathbf{C}$, leading to a global objective function,

$$\frac{1}{2}(\mathbf{x}^{\mathrm{T}}\mathbf{Cx} + \mathbf{y}^{\mathrm{T}}\mathbf{Cy} + \mathbf{z}^{\mathrm{T}}\mathbf{Cz}). \qquad (6\text{-}20)$$

Alternatively, (6-20) can be optimized by solving the following equation,

$$\mathbf{Ci} = \mathbf{f_i}, \forall i \in \{\mathbf{x}, \mathbf{y}, \mathbf{z}\}, \qquad (6\text{-}21)$$

where $\mathbf{f_i}$ is the force vectors in the x, y, and z directions. The applied forces comprise two components: thermal forces and overlap forces. Since the goal is to reduce the temperature, the thermal forces are set equal to the negative of the thermal gradients. This assignment places the blocks far from the high-temperature regions.

After the stiffness matrices are constructed, an initial random placement of the circuit blocks is generated. Based on this placement, the initial forces are computed, permitting the final placement of the blocks to be iteratively determined. This recursive procedure progresses as long as the improvement is above some threshold value and includes the following steps [209]:

1. The power vector resulting from the new placement is determined.
2. The temperature profile of the 3-D stack is calculated.
3. The new value of the thermal and overlap forces are evaluated.
4. The matrices of the repulsive forces are updated.
5. A new placement is generated.

Once the algorithm converges to a final placement, a postprocessing step follows. During this step, the circuit blocks are positioned without any overlaps within the planes of the system. If one plane is packed, the remaining cells that are initially destined for this plane are positioned onto the adjacent planes. A similar process takes place in the y-direction to align the circuit blocks into rows. A divide and conquer method is applied to avoid any overlaps within each row. A final sorting step in the x-direction includes a postprocessing procedure, after which no overlap among cells should exist.

The efficiency of this force-directed placement technique has been evaluated on the MCNC [211] and IBM-PLACE benchmarks [212], demonstrating a 1.3% decrease in the average temperature, a 12% reduction in the maximum temperature, and a 17% reduction in the average thermal gradient. The total wirelength, however, increases by 5.5%. As shown by these results, this technique primarily achieves a uniform temperature distribution across each plane, resulting in a significant decrease in thermal gradients as well as the maximum temperature. The average temperature throughout a 3-D IC, however, is only slightly decreased. This technique focuses on mitigating the hot spots across a multiplane system.

6.3 THERMAL MANAGEMENT TECHNIQUES EMPLOYING THERMAL VIAS

The positive effect of interplane interconnects in decreasing the peak temperature while facilitating the flow of heat in 3-D circuits is demonstrated in Section 6-1. Consequently, interplane vias can conduct heat in addition to propagating a signal. To further enhance the heat transfer process, additional vias that do not function as a signal path can also be utilized. These vias are typically called thermal or dummy vias to emphasize the objective of conveying heat rather than providing signal communication for circuits located on different physical planes.

There are several ways to place thermal vias to decrease the temperature of the upper planes in a 3-D IC. These techniques can include thermal via insertion within certain regions, and dispersion of the thermal vias across the planes with thermal via planning, discussed in Sections 6.3.1 and 6.3.2, respectively. In addition to thermal vias, thermal wires can be employed to transfer heat, as described in Section 6.3.3.

6.3.1 **Region-Constrained Thermal Via Insertion**

In Chapter 5, available space among the cells on each plane is reserved for allocating decoupling capacitors to mitigate simultaneous switching noise. In the same context, the available space can also be used to mitigate thermal problems by allocating thermal vias within these regions. The generation of these regions, however, assumes that sophisticated placement tools can produce free area among the placed cells. Thermal via insertion is applied as a subsequent step before routing the signal nets. Thermal via insertion can satisfy a variety of design objectives (not simultaneously) such as [213]

- maximum or average thermal gradient (g_{max} or g_{ave})
- maximum or average temperature (T_{max} or T_{ave})
- maximum or average thermal via density (d_{thmax} or d_{thave})

The design objective is to identify those regions where thermal vias are most needed (the hot spots) and place thermal vias within those regions at the appropriate density. Such an assignment, however, is mainly restricted by two factors: the routing blockage caused by these vias and the size of the unoccupied regions or white space that exist within each plane. Note that, although the density of the thermal vias can vary among different regions, the thermal vias within each region are uniformly distributed.

In order to determine the required number of thermal vias for a 3-D circuit, the temperature at specific nodes within the volume of a circuit needs to be evaluated. The finite element method based on (6-11)–(6-12) and (6-18)–(6-19) is utilized to determine the temperature of the nodes within a 3-D grid [213]. An iterative approach is applied to determine the thermal conductivity of certain elements, similar to that shown in Figure 6-8 to minimize the thermal objective. More specifically, when initializing the optimization procedure, an ideal value, which depends on the desired objective as well as the initial temperature profile, is determined to characterize the thermal gradients within each of the thermal via regions. In addition, the thermal conductivity of these regions is set to the minimum value, which coincides with no thermal vias in these regions. The thermal conductivity of the regions is iteratively modified, updating the temperature of the nodes. The algorithm terminates when any further change in the thermal conductivity does not significantly improve the desired objective.

Inserting thermal vias can significantly affect the thermal conductivity. Since these vias facilitate the transfer of heat in the z-direction, the thermal conductivity in the z-direction greatly differs from the conductivity in the x- and y-directions. To quantify the effect of the thermal vias in terms of a change in the thermal conductivity within a 3-D grid, the following expressions can be used to determine the change in the global stiffness matrix in (6-19):

$$K_z^{eff} = d_{th}K_{via} + (1 - d_{th})K_z^{plane}, \tag{6-22}$$

$$K_x^{eff} = K_y^{eff} = \left(1 - \sqrt{d_{th}}\right)K_{lateral}^{plane} + \frac{\sqrt{d_{th}}}{\frac{1-\sqrt{d_{th}}}{K_{lateral}^{plane}} + \frac{\sqrt{d_{th}}}{K_{via}}}. \tag{6-23}$$

d_{th} is the density of the thermal vias and $K_{lateral}^{plane}$ and K_z^{plane} are the thermal conductivity of a physical plane of a 3-D system in the horizontal and vertical directions, respectively, where no thermal vias are employed. These values are 2.15 W/m-K and 1.11 W/m-K, respectively [213], for the MITLL 3-D process technology [140]. K_{via} is the thermal conductivity of the vias, which is equal to the thermal conductivity of copper, 398 W/m-K. With these expressions, the vertical and lateral thermal conductivity or, alternatively, the density of the thermal vias in each region is iteratively determined until the specified thermal gradient or temperature, according to the specified objective, is reached. Note that (6-22) is identical to (6-8) in the simple thermal model based on analytic expressions.

In Figure 6-13, (6-22)–(6-23) are plotted for a variable density of thermal vias. The thermal conductivity in the z-direction is about two orders of magnitude greater than in the horizontal direction. Therefore, only the vertical thermal conductivity is updated during the optimization process, while the thermal conductivity in the horizontal directions can be determined from the new thermal via density obtained from the algorithm and (6-23).

This method has been applied to the MCNC and IBM-PLACE benchmarks, where the interconnect power consumption is not considered.

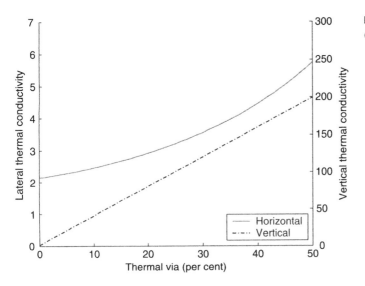

■ **FIGURE 6-13** Thermal conductivity vs. thermal via density [213].

Table 6-4 Average percent change of various thermal objectives for the case with no thermal vias [213].

Objective	Average Percent Change					
	g_{max}	g_{ave}	T_{max}	T_{ave}	d_{thmax}	d_{thave}
g_{max}	-68.1	-60.8	-44.5	-25.9	44.9	10.2
g_{ave}	-75.7	-70.7	-51.6	-29.5	50	17.6
T_{max}	-71.1	-64.5	-47.3	-27.3	50	12.3
T_{ave}	-73.2	-67.4	-49.2	-28.3	50	14.3
d_{thmax}	-55.5	-43.3	-31.4	-19.2	25	4.2
d_{thave}	-79.2	-75.3	-54.7	-31.0	50	23.9

Some of these results are listed in Table 6-4. Note that the target objectives are decreased by increasing the density of the thermal vias. A considerable reduction in all of the aforementioned objectives (see p. 121) is observed. In addition, as compared to a uniform distribution of thermal vias, fewer thermal vias are inserted to reduce the temperature. Furthermore, by analyzing the distribution of the thermal vias throughout a 3-D circuit consisting of four planes, the density of the vias is smaller in the upper planes. This behavior can be explained by noting that in a 3-D IC, the thermal gradients can substantially increase the temperature in the upper planes. By placing additional thermal vias in the lower planes, the thermal gradients are mitigated, reducing the temperature on the upper planes.

6.3.2 **Thermal Via Planning Techniques**

Although thermal vias can considerably facilitate the flow of heat toward the ambient, the placement of the vias can adversely affect other design objectives, such as the area and wirelength of a 3-D circuit. The density of the thermal vias is, therefore, limited and cannot arbitrarily increase. Consequently, thermal via planning can be described as the problem of minimizing the number of thermal vias while the temperature of the circuit and the capacity of the thermal vias are constrained. Compact thermal models, such as the model described in subsection 6.1.2, are preferred due to the lower computational time required to obtain the thermal profile of 3-D circuits as compared to finite element and finite difference methods [203], [214]–[216]. Using as a baseline, the compact thermal model described in subsection 6.1.2, a 3-D circuit is discretized into tiles. The tiles located on the same x-y coordinates but on different planes constitute a pillar modeled by a group of serially connected resistors and heat sources (see Figure 6-6).

Based on this model, the problem of determining the minimum number of through silicon vias (TSVs) necessary to satisfy a specific temperature constraint can be described by the following nonlinear programming (NLP) problem:

$$\min \sum_{i=2}^{n} d_{thi}, \qquad (6\text{-}24)$$

where d_{thi} is the thermal TSV density on plane i for a 3-D circuit comprised of n physical planes. In addition, a number of constraints apply to (6-24), such as temperature constraints (*i.e.*, the temperature of the circuits cannot exceed a specified value), capacity constraints for the TSVs in each tile, a lower bound constraint for the number of TSVs to ensure that the wirelength of the circuit does not increase due to an insufficient number of signal TSVs in a specific tile, and a heat flow equality constraint (*i.e.*, the incoming and outgoing heat flow for every tile should be equal).

Relating the thermal conductivity of the TSVs with the serially connected resistors in a single pillar, (6-24) can be rewritten as

$$\text{min.} \quad \sum_{k \geq 2} \left(\frac{R_{via} I_{i,j,k}}{T_{i,j,k} - T_{i,j,k-1}} - n_{TSV} \right), \qquad (6\text{-}25)$$

where R_{via} is the thermal resistance of one TSV [196] and n_{TSV} is the number of TSVs, which is equivalent to the thermal resistance of a tile within a 3-D grid. I and T are the heat flow in the z-direction and temperature of the tiles in the grid, respectively, and i, j, and k are the index of the tiles.

Efficiently solving this NLP is a formidable task. The thermal via planning process is therefore divided into a two-stage problem: determining the intraplane thermal TSV density within each plane of a 3-D circuit and determining the interplane thermal TSV density among the planes within the stack. Depending on the formulation of these problems and the applied constraints, different intraplane and interplane distributions of the thermal TSVs are obtained.

The technique of multilevel routing [216], [217] is extended to 3-D ICs, including thermal via planning. Multilevel routing with thermal via planning can be treated as a three-stage process, illustrated in Figure 6-14: a coarsening phase, an initial solution generation phase at the coarsest level of the grid, and a subsequent refinement process phase until the finest level of the grid is reached. Before the coarsening phase is initiated, the routing resources, the capacity of the TSVs, and the power density in each tile are determined. The power density and routing resources are determined at each coarsening step. At the coarsest level (level k), an initial routing tree is generated. At this point, the TSV planning step is invoked, assigning TSVs to each tile within a coarse grid. During the refinement phase, the TSVs are distributed to preserve the solution produced at the previous level. If the final temperature at the end of the refinement phase does not satisfy a specified temperature, the TSVs are further adjusted to satisfy the target temperatures.

The TSV planning step is based on the alternating direction TSV planning algorithm (ADVP), which distributes the TSVs in alternate directions: in the first step, among the planes of the 3-D IC, and in the second step, within each plane of the circuit. This algorithm aims at reducing the overall runtime as the thermal profile required during the various stages of the multilevel routing process can considerably

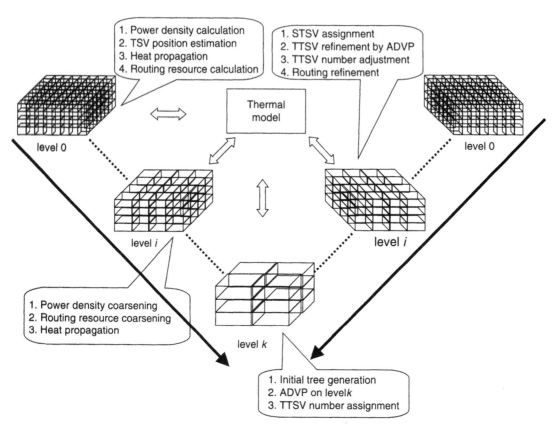

■**FIGURE 6-14** Multilevel routing flow with thermal via planning [215].

increase the execution time. The problem of distributing the TSVs among the planes of a 3-D system can be described as a convex problem. An analytic solution is determined if the capacity bounds for the TSVs are removed,

$$
a_n : a_{n-1} : \ldots : a_3 : a_2 = \sqrt{P'_n} : \sqrt{P'_n + P'_{n-1}} : \ldots : \sqrt{\sum_{k=3}^{n} P'_k} : \sqrt{\sum_{k=2}^{n} P'_k},
$$

$$(6\text{-}26)$$

where a_i and P'_i are, respectively, the number of TSVs and the power density of each tile for a grid consisting of n planes.

A corresponding analytic solution cannot be easily determined, however, for the horizontal or intraplane distribution of TSVs. Alternatively, *heat propagation* and *path counting* replace the thermal profiling step. Heat propagation considers the propagation of the

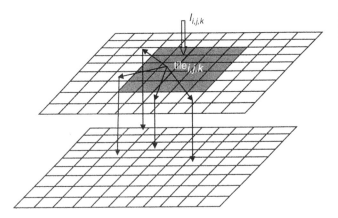

■ **FIGURE** 6-15 Various heat propagation paths within a 3-D grid [215].

heat flow among the tiles of the grid and is determined by evaluating the different paths for transferring heat to the lower planes of the grid. Different heat propagation paths are illustrated in Figure 6-15.

The multilevel routing and ADVP algorithm are applied to MCNC benchmarks and compared both to the TSV-planning approach presented in [214] and to a uniform distribution of TSVs. As listed in Table 6-5, the ADVP algorithm achieves a significant decrease in the required number of TSVs in order to maintain the same temperature, possibly resulting in a lower fabrication cost and less routing congestion. From Table 6-5, a considerable reduction in the number of TSVs is achieved by the ADVP algorithm without any increase in computational time.

Although the ADVP algorithm reduces the required amount of thermal TSVs, a further decrease in the density of the thermal TSVs can be achieved by applying a different approach to determining the TSVs densities. With this approach, both the interplane and intraplane via density problem is converted to a convex problem through several simplifications [218]. The primary assumption of this approach is that the silicon substrate of each of the upper physical planes and the bonding material at the interface between two planes are treated as a single material with homogeneous material properties. The thickness of this material is the summation of the thickness of the silicon layer and the bonding layer, with a thermal conductivity K_{avg} equal to the average thermal conductivity of the silicon and bonding material. The solution of these convex problems produces a different distribution of interplane and intraplane TSVs [218], as compared to the distribution produced by ADVP and a uniform TSV distribution. These differences are summarized in Table 6-6, where K_{via} is the thermal conductivity of the TSVs and S is the area of the circuit blocks within each plane of a 3-D system.

Table 6.5 Comparison of TSV planning techniques.

Circuits	T [°C]	[215] m-ADVP # TSV	Area Ratio	Runtime [sec]	T[°C]	[214] m-VPPT # TSV	Area Ratio	Runtime [sec]	T [°C]	Uniform # TSV	TSV Area Ratio	Runtime [sec]
ami33	77.0	1282	2.5%	1.55	77.1	1801	3.5%	1.76	77.1	2315	4.5%	1.62
ami49	77.0	20956	0.9%	13.5	77.1	43794	1.8%	12.15	76.9	166366	6.8%	16.17
n100	77.0	11885	1.5%	7.66	77.0	22211	2.8%	8.31	76.8	30853	3.9%	7.54
n200	77.0	13980	1.8%	12.24	77.2	18835	2.4%	10.89	77.1	30346	3.9%	12.21
n300	77.0	17646	1.3%	20.44	77.1	30161	2.2%	21.73	76.9	57342	4.2%	22.42
Avg.		1.0	1.6%	1.0		1.68	2.6%	1.01		3.55	4.6%	1.06

Table 6-6 Different solutions for distributing TSVs in 3-D ICs.

Algorithm	Interplane Planning	Intraplane Planning
m-VPPT [232]	$d_{thi} : d_{thj} = l_i : l_j$	$d_{thik} : d_{this} = l_{ik} : l_{is}$
m-ADVP [233]	$\frac{d_{thi}+\alpha}{d_{thj}+\alpha} = \sqrt{l_i} : \sqrt{l_j}, \alpha = \frac{K_{via}}{(K_{avg}S)}$	$d_{thik} : d_{this} = l_{ik} : l_{is}$
TVP [236]	$\frac{\lambda d_{thi}+1}{\lambda d_{thj}+1} = \sqrt{l_i} : \sqrt{l_j}, \lambda = \frac{K_{via}}{K_{avg}} - 1$	$d_{thik} : d_{this} = \sqrt{l_{ik}} : \sqrt{l_{is}}$

This modified thermal via planning step has been applied to MCNC and GSRC benchmark circuits and compared with the ADVP and other TSV distribution algorithms [214], [215]. Some results are reported in Table 6-7, where the solution of the approximate convex problem reduces the number of thermal vias required to reach a pre-specified temperature.

The improved thermal via planning step is integrated into a hierarchical floorplanning technique [218] for 3-D circuits where the circuit blocks are initially partitioned onto the planes of the circuit. Since no interplane moves are allowed after the partitioning step is completed, the partitioning step is crucial in determining the overall quality of the final result. This partitioning problem is approached as a sequence of knapsack problems [219], where the hottest blocks are placed on the lower planes of a 3-D circuit to prevent steep thermal gradients as the heat generated by these blocks is conveyed to the heat sink. Furthermore, overlap (*i.e.*, placement of the blocks on adjacent physical planes with the same *x-y* coordinates) among these high-

Table 6-7 Comparison among the required number of TSVs.

Circuit	m-ADVP [215]		m-VPPT [214]		TVP [218]	
	T_{max}	# T-via	T_{max}	# T-via	T_{max}	# T-via
ami33	76.8	1109	76.7	1360	77.5	995
ami49	77.0	21668	77.1	28793	77.2	20310
n100	77.2	16731	76.9	25205	77.0	14533
n200	77.1	14273	76.4	17552	77.1	12869
n300	76.8	19337	76.5	25995	76.9	18614
Avg.		1.12		1.51		1.02

power density blocks is avoided. The integrated thermal via planning and floorplanning approach is compared with a nonintegrated approach, where the floorplan (initially omitting the thermal objectives) is generated first followed by thermal via planning as a postprocessing step. The integrated technique requires 16% fewer thermal vias for the same temperature constraint, with a 21% increase in computational time and an almost 3% reduction in the total area.

6.3.3 **Thermal Wire Insertion**

In addition to the benefits that the added thermal vias produce, thermal wires can also be utilized to enhance the heat transfer process. These thermal wires correspond to horizontal wires that connect regions with different thermal via densities through thermal interplane vias. These thermal wires are treated as routing channels wherever there are available tracks. Both thermal interplane vias and wires can be handled during routing [220]. Given a placement of cells within a 3-D IC, the technology parameters, and a temperature constraint, sensitivity analysis and linear programming methods are utilized to route a circuit. For routing purposes, a 3-D grid is imposed on a 3-D circuit as shown in Figure 6-16 for a two-plane 3-D circuit. In addition, the thermal model of a circuit is based on a resistive network, as discussed in subsection 6.1.2. Note that the interconnect power component is not considered in this thermal model [220].

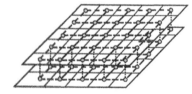

■ FIGURE 6-16 Routing grid for a two-plane 3-D IC. Each horizontal edge of the grid is associated with a horizontal wire capacity. Each vertical edge is associated with an interplane via capacity.

Placing thermal vias and wires to decrease the circuit temperature adversely affects the available routing resources while increasing the routing congestion. Each vertical edge of the routing grid is, therefore, associated with a specific capacity of interplane vias. A similar constraint applies for the horizontal edges, which represent horizontal routing channels. The width of the routing channel can be considered equal to the edge width of the tiles. As shown in Figure 6-17, the thermal wire and vias affect the routing capacity of each tile.

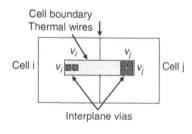

■ FIGURE 6-17 Impact of a thermal wire on the routing capacity of each grid cell. v_i and v_j denote the capacity of the interplane vias for cell i and j, respectively. The horizontal cell capacity is equal to the width of the boundary of the cells [220].

The 3-D global routing flow is depicted in Figure 6-18 where thermal vias and wires are inserted to achieve a target temperature under congestion and capacity constraints. A 3-D minimum Steiner tree is initially generated, followed by an interplane via assignment. A 2-D maze router produces a thermally driven route within each plane of the circuit. In the following steps [220], an iterative procedure is applied to insert the thermal vias and wires and complete the physical routing. A thermal model and sensitivity analysis are used to perform a linear programming based thermal via and wire insertion in the first

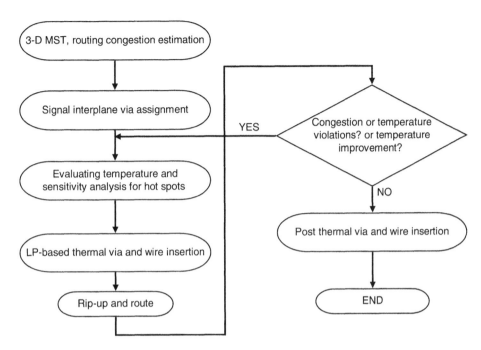

■ **FIGURE 6-18** Flowchart of a temperature aware 3-D global routing technique [220].

step of the iterative procedure. If this insertion violates the congestion constraints or causes overflow in the routing channels, a rip-up and route step is used to resolve these conflicts [220]. Thermal profiling and sensitivity analysis are repeated to determine whether the target temperature has been achieved. The iterations are terminated when there are no remaining violations or the temperature of the circuit cannot be further improved. The complexity of the algorithm is bounded by the complexity of the iterative procedure. The complexity of each iteration is $O(NGlogG+G^3)$, where N and G are the number of nets in the circuit and the number of cells in the routing grid, respectively.

6.4 **SUMMARY**

Thermal analysis and related design techniques for 3-D ICs are discussed in this chapter. A summary of these techniques follows.

■ High temperatures deteriorate circuit reliability and lifetime, increase leakage power, and decrease circuit performance.

- Higher temperatures and thermal gradients are predicted for 3-D ICs due to increased power densities and greater distances between the circuits on the upper planes and the heat sink.

- Thermal coupling can further increase the self-heating of devices, leading to hot spots.

- Thicker silicon substrates facilitate the heat transfer process; however, high aspect ratio interplane vias reduce the performance of the circuit and are difficult to fabricate.

- Thermal models of 3-D circuits include analytic expressions, compact thermal resistive networks, and 3-D grids for finite element analysis, in ascending order of increasing computational complexity and accuracy.

- The increase in temperature of a 3-D circuit exhibits a square and linear relationship with the number of planes and the power density of each plane, respectively.

- The boundary conditions for the thermal model of a 3-D circuit typically assume adiabatic walls for the lateral and top surfaces and an isothermal surface for the bottom surface attached to the heat sink.

- Physical design techniques, such as floorplanning, placement, and routing, that embody a thermal objective can be a useful tool to manage thermal issues in 3-D ICs.

- Design techniques can reduce thermal gradients and temperatures in 3-D circuits by redistributing the blocks among and within the planes of a 3-D circuit.

- The thermal-electric duality can be exploited to apply well established circuit analysis techniques, such as the application of modified nodal analysis, to thermal design techniques.

- The computational time of the thermal management techniques depends on the accuracy of the thermal model. More accurate thermal models achieve improved results but can require unacceptably high computational times. Alternatively, simpler models reduce runtime but can provide insufficient accuracy.

- A compromise between the accuracy and the computational requirements of the thermal model is necessary.

- The heat primarily spreads vertically towards the heat sink rather than laterally. Consequently, the thermal resistance in the horizontal direction can be considered as constant to improve the computational time for thermal profiling.

- The third dimension can greatly increase the computational time by significantly enlarging the solution space of the thermal design techniques.

- Interplane vias that do not carry an electrical signal are called thermal or dummy vias. These thermal vias can be utilized in 3-D circuits to convey heat to the heat sink.

- Dummy vias create routing obstacles. These vias should therefore be judiciously inserted.

- Thermal via planning describes the problem of minimizing the number of thermal vias while satisfying temperature and interplane via capacity constraints. Thermal via planning techniques can significantly decrease the temperature and thermal gradients within a 3-D circuit.

- Thermal wires in the horizontal direction are equivalent to thermal vias and can be utilized to lower thermal gradients in 3-D circuits.

7

Timing Optimization for Two-Terminal Interconnects

The 3-D interconnect prediction models discussed in Chapter 4 indicate a considerable reduction in interconnect length for 3-D circuits, which can improve the speed and power characteristics of modern integrated circuits. In order to exploit this advantage, numerous physical design techniques for 3-D circuits satisfying a variety of design objectives, such as area, wirelength, routing congestion, and temperature, have been developed. In addition to this important advantage, three-dimensional integration demonstrates many opportunities for heterogeneous SoCs [138].

Integrating circuits from diverse fabrication processes into a single-multiplane system can result, however, in substantially different impedance characteristics for the interconnects of each physical plane within a 3-D circuit. These particular interconnect traits present new challenges and opportunities for 3-D circuits. This situation is further complicated by the interplane through silicon vias, which can exhibit different impedance characteristics as compared to the horizontal (or intraplane) interconnects. By considering the disparate interconnect impedance characteristics of 3-D circuits, the performance of the interplane interconnects can be significantly improved. As suggested by the interconnect prediction models discussed in Chapter 4, these interplane interconnects are typically the longest lines within a 3-D circuit. Improving the performance of these global interconnects is therefore of significant importance, since the overall performance of the system is also significantly enhanced. In this chapter, a technique and algorithm to decrease the delay of two-terminal interplane interconnects by optimally placing the interplane vias are described. In the following section, the characteristics of the interplane interconnects which require a different interconnect model as compared to 2-D circuits are discussed.

A closed-form solution for the minimum delay of a two-terminal interplane net that includes only one interplane via is provided in Section 7.2. The performance degradation that can occur by not optimally placing the interplane via is also discussed. Two-terminal nets that comprise more than one interplane via are investigated in Section 7.3. In addition, an algorithm for placing interplane vias is described along with simulation and analytic results. A comparison of the proposed technique with a wire sizing algorithm is also presented. A short summary of the chapter is presented in the last section.

7.1 INTERPLANE INTERCONNECT MODELS

The impedance characteristics of the metal layers that belong to different physical planes of a 3-D circuit can vary significantly. This variation can be attributed to several causes. For example, consider a 3-D circuit in which a processor is integrated with a few memory planes. The fabrication process used for the processor and the memory plane can result in different impedances among the metal layers of the bottom and upper planes. In addition, the magnitude of the process variations, which can affect, for example, the metal layer and interlayer dielectric thickness of each of those processes is different. Consequently, the nominal impedance of the lines in each plane is affected differently. As a more specific example, consider the fabrication process for 3-D circuits manufactured by the MITLL [140], [221]. In this process, the third plane of the 3-D circuit can be optimized for RF circuits. In this case, the resistance of the metal layers in the RF plane is approximately an order of magnitude lower than that of the metal line resistance in the other digital planes.

If the same manufacturing process, however, is used to fabricate all of the planes of a 3-D circuit, variations in the impedance characteristics of the lines in a 3-D circuit can also originate from other causes, such as the bonding style and technology. Consider the interconnect structure, which is extracted to determine the capacitance on the topmost metal layer, shown in Figures 7-1a and 7-1b for a 2-D and 3-C circuit, respectively [222]. In a 2-D circuit, the metal layer located immediately below the parallel lines is treated as a ground plane. In a 3-D circuit where front-to-front bonding is employed, the topmost metal layer of the second plane, which is flipped and bonded onto the first plane, behaves as a second ground plane for the topmost metal layer of the first physical plane of the 3-D circuit. Alternatively, in the case of front-to-back bonding, a ground plane can result from the substrate of the second plane if bulk CMOS technology is used to fabricate the planes of the 3-D circuit. This second ground plane can significantly alter the capacitance of the interconnects. As an example, consider the MITLL 3-D process, where the 3-D system is illustrated in Figure 7-2. In this technology, the lines of the topmost metal layer of the first and second plane exhibit a different capacitance as compared to the capacitance of the global metal layer on the third plane.

Another important factor in determining the performance of the interplane interconnects is the impedance characteristics of the interplane vias, which depend on the bonding style and technology.

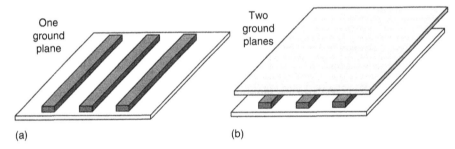

■ **FIGURE 7-1** Global interconnect structures for impedance extraction: (a) three parallel metal lines over a ground plane in a 2-D circuit and (b) three parallel metal lines sandwiched between two ground planes in a 3-D circuit.

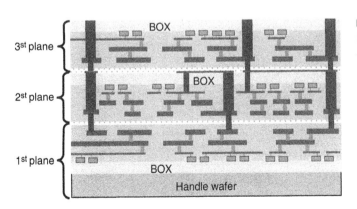

■ **FIGURE 7-2** A three plane FDSOI 3-D circuit [140], [221]. Planes one and two are front-to-front bonded, while planes two and three are front-to-back bonded.

Front-to-back bonding will likely result in interplane vias with a greater length as compared to front-to-front bonding since in the latter case the vertical distance among the planes is greater. Furthermore, the type of technology can have a profound effect on the impedance of the interplane vias.

To justify this argument, the capacitance of different interplane via structures illustrated in Figures 7-3 to 7-5 has been extracted with a commercial impedance extraction tool [224]. The structure shown

■ **FIGURE 7-3** Capacitance extraction for an interplane via structure: (a) interplane via surrounded by orthogonal metal layers and (b) capacitance values for several via sizes and spacings.

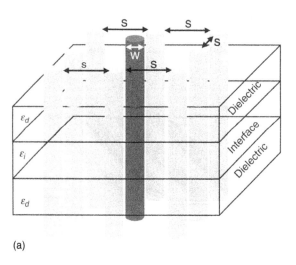

(a)

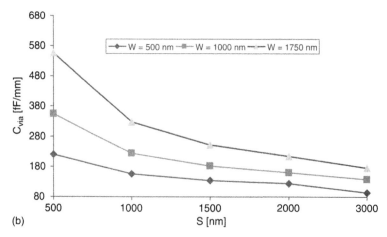

(b)

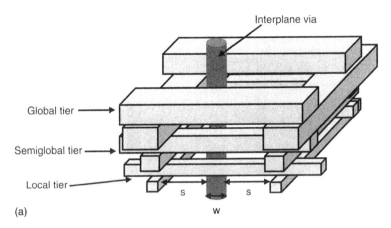

Interplane via

Global tier

Semiglobal tier

Local tier

S   S

w

(a)

■ **FIGURE 7-4** Capacitance extraction for an interplane via structure: (a) interplane via through layers of dielectric and the bonding interface, surrounded by eight interplane vias and (b) capacitance for various via sizes and spacings. For all of the layers, the same dielectric material is assumed (*i.e.*, $\varepsilon_d = \varepsilon_1 = \varepsilon_{SiO_2}$).

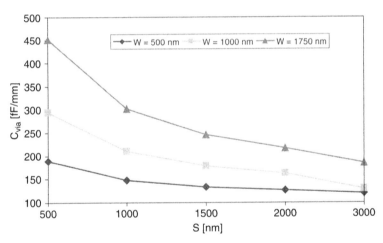

in Figure 7-3 corresponds to that segment of the interplane via surrounded by the horizontal metal levels of a plane within a 3-D system and is independent of the target technology (*i.e.*, bulk CMOS or SOI). The number of metal levels surrounding the via, however, depends on the fabrication process. Alternatively, the structure shown in Figure 7-4 corresponds to an interplane via surrounded by other vias, which traverses the layers of the dielectric and bonding material (assuming, for simplicity, identical dielectric constants). Such a via structure can exist in an SOI technology. For 3-D circuits where each plane is fabricated with a mainstream

■ **FIGURE 7-5** Capacitance extraction for an interplane via structure: (a) interplane via through silicon substrate, surrounded by a thin insulator layer and (b) capacitance for various via sizes and thicknesses of the insulator layer.

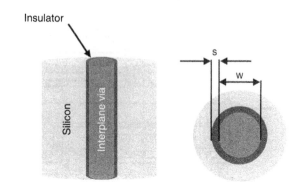

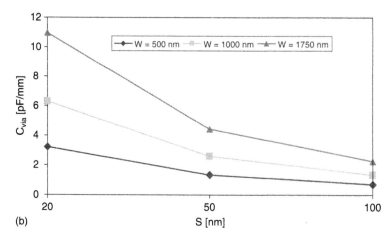

(b)

CMOS process or with a combination of SOI and CMOS processes, the structure in Figure 7-5 can be utilized for capacitance extraction.

The interplane interconnects are therefore modeled as an assembly of horizontal interconnect segments with different impedance characteristics connected by interplane vias. This model is considerably different from the typical interconnect model used in a 2-D circuit, where a two-terminal net is modeled by a single segment with the same uniform distributed impedance throughout the interconnect length. An interplane interconnect connecting two circuits located on different physical planes is illustrated in Figure 7-6. In the following sections, it is shown that the delay of this type of 3-D interconnect can be improved by appropriately placing the interplane vias.

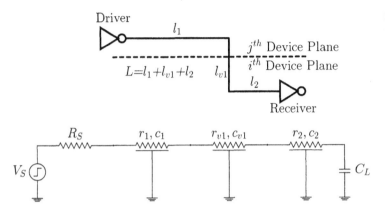

7.2 TWO-TERMINAL NETS WITH A SINGLE-INTERPLANE VIA

In this section, the simple case of an interplane two-terminal net connecting two circuits located on different physical planes including only one interplane via is considered. The delay model and the dependence of the delay on the via location are discussed in subsection 7.2.1. A physical explanation of the variation in delay with the via location is provided in subsection 7.2.2, whereas the optimum via location for different impedance characteristics of the interconnect segments is determined in subsection 7.2.3. The improvement in delay for this type of interconnect is demonstrated in subsection 7.2.4.

7.2.1 Elmore Delay Model of an Interplane Interconnect

As previously mentioned, due to the nonuniformity of the interconnects, each segment is modeled as a distributed RC line with different impedance characteristics, as shown in Figure 7-6. The driver is modeled as a step input voltage source with a linear output resistance R_S, and the interconnect is terminated with a capacitive load C_L. The total resistance and capacitance of the horizontal segments 1 and 2 are $R_1 = r_1 l_1$, $R_2 = r_2 l_2$, and $C_1 = c_1 l_1$, $C_2 = c_2 l_2$, where r_1, r_2, and c_1, c_2 denote the resistance and capacitance, respectively, per unit length. The length of the two horizontal segments is l_1 and l_2, and the via length is l_{v1}. Since the via may span more than one physical plane, l_{v1} can be expressed as

$$l_{v1} = (n-1)l_v, \tag{7-1}$$

where n is the number of physical planes among the connected circuits and l_v is the length of the via connecting two metal layers located on two adjacent physical planes. This via length is determined by the fabrication process and can typically range from 10 μm to 200 μm [38], [107], [225]. The total length of the line can be expressed as

$$L = l_1 + l_{v1} + l_2. \tag{7-2}$$

Due to the nonuniform impedance characteristics of the line, the via location or, alternatively, the length of each horizontal wire segment affects the delay of the line. Thus, the objective is to place the via such that the interconnect delay is minimum. To analyze the delay of a line, the distributed Elmore delay model has been adopted due to the simplicity and high fidelity of this model [226]. The accuracy of the model can be further improved as discussed in [227]. However, unlike a single plane, more than one set of fitting coefficients is required in a 3-D system. The objective is to determine the via location or, alternatively, the length of segments 1 and 2, that minimizes the distributed Elmore delay of the line.

The distributed Elmore delay for the system shown in Figure 7-6 is

$$T_{el} = \frac{R_1 C_1}{2} + C_1 R_S + \frac{R_{v1} C_{v1}}{2} + C_{v1}(R_S + R_1) + \frac{R_2 C_2}{2} \tag{7-3}$$
$$+ C_2(R_S + R_1 + R_{v1}) + C_L(R_S + R_1 + R_{v1} + R_2).$$

Substituting the total resistance and capacitance with the per unit length parameters, and using (7-1) and (7-2), the Elmore delay described in (7-3) is a function of the length of the first segment l_1,

$$T_{el}(l_1) = A_1 l_1^2 + A_2 l_1 + A_3, \tag{7-4}$$

where

$$A_1 = (1/2)(r_1 c_1 - 2r_1 c_2 + r_2 c_2), \tag{7-5}$$

$$A_2 = R_s(c_1 - c_2) + (n - 1)l_v(r_1 c_{v1} - r_{v1} c_2 + r_2 c_2 - r_1 c_2) + L(r_1 c_2 - r_2 c_2) + C_L(r_1 - r_2), \tag{7-6}$$

$$A_3 = ((n - 1)l_v)^2 \left(\frac{r_{v1} c_{v1}}{2} - r_{v1} c_2 + \frac{r_2 c_2}{2} \right)$$
$$+ (n - 1)l_v((Lc_2 + C_L)(r_{v1} - r_2) + R_S(c_{v1} - c_2))$$
$$+ C_L(R_S + Lr_2) + R_S Lc_2 + \frac{L^2 r_2 c_2}{2}. \tag{7-7}$$

Equation (7-4) describes a parabola, but the convexity is not guaranteed since A_1 can be negative. The second derivative of (7-4) with respect to l_1 is

$$\frac{d^2 T_{el}}{dl_1^2} = 2A_1. \tag{7-8}$$

Depending on the sign of A_1, the propagation delay of the line exhibits either a minimum or a maximum as l_1 varies or, alternatively, as the location of the via along the line changes. The following notations are introduced to facilitate the analysis,

$$r_{21} = \frac{r_2}{r_1}, \tag{7-9}$$

$$c_{12} = \frac{c_1}{c_2}. \tag{7-10}$$

From (7-9) and (7-10), the second derivative is

$$\frac{d^2 T_{el}}{dl_1^2} = r_1 c_2 (r_{21} + c_{12} - 2). \tag{7-11}$$

Since $r_1 c_2$ is always positive, the sign of (7-11) and, consequently, the timing behavior of the line only depend on the sign of the term in the parentheses. For the propagation delay to be minimum, the following inequality should be satisfied,

$$r_{21} + c_{12} - 2 > 0. \tag{7-12}$$

Note that the notations introduced in (7-9) and (7-10) describe resistance and capacitance ratios and, consequently, are dimensionless. If (7-12) is not satisfied, the delay of the line exhibits a maximum.

7.2.2 Interplane Interconnect Delay

In the previous subsection, (7-12) only depends on the impedance characteristics of the segments that constitute the line. The dependence of the delay on the impedance of the segments of the line is analyzed in this subsection. To better explain this dependence, consider initially an *RC* interconnect line of uniform impedance. An optimum wire-shaping function of the propagation delay has been shown to exist for *RC* interconnects [228]. The optimum wire sizing function is a monotonically decreasing function of the width, as illustrated in Figure 7-7. This decrease in width occurs because close to the receiver the downstream capacitance is small, and, therefore, this capacitance is charged by a larger resistance that results from the reduction in line width. Alternatively, the downstream capacitance close to the driver is large, and, consequently, a small resistance charges the capacitance, due to the widening of the interconnect toward the driver.

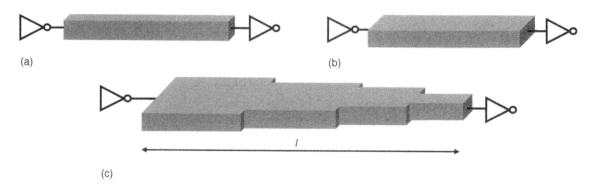

(a)   (b)

(c)

■ **FIGURE 7-7** An example of interconnect sizing. (a) An interconnect of minimum width, W_{min}, (b) uniform interconnect sizing $W > W_{min}$, and (c) nonuniform interconnect sizing $W = f(l)$.

An interplane interconnect with nonuniform impedance characteristics can be treated as a tapered uniform line consisting of only two segments. If r_0 and c_0 denote the resistance and capacitance per unit area, respectively, the resistance and capacitance per unit length of the segments of the interplane interconnect can be described as $r_1 = r_0/w_1$, $r_2 = r_0/w_2$, $c_1 = c_0 w_1$, and $c_2 = c_0 w_2$, where w_1 and w_2 are the width of the segments of the line. Thus, r_{21} and c_{12} characterize the tapering factor of the line. If $r_{21} > 1$ and $c_{12} > 1$, the tapering factor is smaller than one, and the delay of the line can be reduced by appropriately selecting the length of the line segments or, alternatively, the via location. These values of the impedance ratios r_{21} and c_{12} correspond to a decreasing shaping function of the interconnect, which decreases the interconnect delay [228]. Consequently, the via location should be selected to ensure that the shape of the interplane interconnect approaches the sizing function of a uniform interconnect comprised of only two segments. Note that the optimum placement of the via does not necessarily achieve the optimum tapering factor as described by the optimum wire sizing function, yet a significant reduction in the delay of the line is possible. In addition, as discussed in subsection 7.3.3, the optimum via placement can achieve a greater improvement in delay as compared to wire sizing for interplane interconnects. This behavior is mainly due to the nonuniform characteristics of the interconnect and the discontinuities that the interplane vias create in sizing the interconnect.

If $r_{21} > 1$ and $c_{12} < 1$, both the resistance and the capacitance per unit length of segment 2 are greater than the RC impedance of segment 1, and, typically, the optimum via location is such that the length of the segment with the higher impedance is minimized. The same relationship applies to the case where $r_{21} < 1$ and $c_{12} > 1$. If $r_{21} < 1$ and $c_{12} < 1$, a longer l_1 increases the resistance of the line, while a longer l_2 increases the capacitance of the line. The tapering factor becomes greater than one, and the delay of the line exhibits a maximum. Consequently, the nonuniform impedance characteristics of the interplane interconnects, which is further enhanced by the impedance of the interplane vias, does not allow the process of placing a via to be treated as optimizing a convex function, which has been demonstrated for wire sizing [229]. This argument is analytically proven in Section 7.3.

7.2.3 **Optimum Via Location**

In subsection 7.2.1, the timing behavior of the line is shown to depend on the impedance characteristics of the segments that comprise the line (see expression (7-12)). Placing a via to minimize the delay based on the impedance of the interconnect segments is presented in this subsection. Note that while the timing behavior of the line with respect to the via location does not depend on the interconnect length, the via location that achieves the minimum delay depends on the interconnect length.

From (7-4)–(7-7), the value of l_1 for which the delay exhibits an extremum, either minimum or maximum, is

$$l_1 = -\left[\frac{(n-1)l_v(r_1c_{v1} - r_{v1}c_2 + r_2c_2 - r_1c_2) + R_s(c_1 - c_2) + L(r_1c_2 - r_2c_2) + C_L(r_1 - r_2)}{r_1c_1 - 2r_1c_2 + r_2c_2}\right].$$

$$(7\text{-}13)$$

Since no constraints have been applied on the value of l_1, the extreme point can occur for values other than within the physical domain of l_1, that is, $l_1 \in [l_d, l_r]$ where $l_d = l_{min}$ and $l_r = L - l_{v1} - l_{min}$. l_{min} determines the minimum distance between a via and a cell, as constrained by the design rules of the fabrication process. The Lemma in Appendix B characterizes the optimum via location for various values of r_{21}, c_{12}, and l_1.

Depending on the sign of (7-11) and the value of l_1 in (7-13), the optimum via location is determined for each possible case:

A) $\frac{d^2 T_{pd}}{dl_1^2} > 0$. If $l_1 \in [l_d, l_r]$, the propagation delay is minimum when l_1 is the value described in (7-13). Consequently, the via should be placed at a distance l_1 from the driver. SPICE measurements of the 50% propagation delay for a 600-μm line are illustrated in Figure 7-8 versus the via location l_1. Two observations can be made. First, that the delay

■ **FIGURE 7-8** SPICE measurements of 50% propagation delay of a 600-μm line versus the via location l_1 for various values of r_{21}. The interconnect parameters are $r_1 = 79.5$ Ω/mm, $r_{v1} = 5.7$ Ω/mm, $c_{v1} = 6$ pF/mm, $c_2 = 439$ fF/mm, $c_{12} = 1.45$, $l_v = 20$ μm, and $n = 2$. The driver resistance and load capacitance are $R_S = 50$ Ω and $C_L = 50$ fF, respectively.

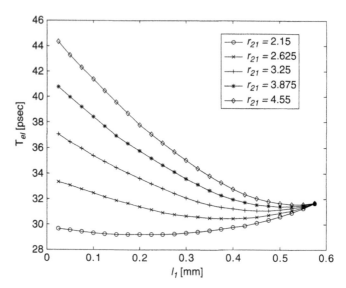

exhibits a minimum and that the minima shift to the right as r_{21} increases. If $l_1 < l_d$, the via should be placed closest to the driver, while if $l_1 > l_r$, the via should be placed closest to the receiver.

B) $\frac{d^2 T_{el}}{d l_1^2} < 0$. In this case, the delay of the line reaches a maximum for the value of l_1, as described in (7-13). If $l_1 \in [l_d, l_r]$, according to the Lemma in Appendix B, for $l_1 < (l_r + l_d)/2$, the via should be placed closest to the receiver, while for $l_1 > (l_r + l_d)/2$, the via should be placed closest to the driver. In Figure 7-9, SPICE measurements of the 50% propagation delay

■ **FIGURE 7-9** SPICE measurements of the 50% propagation delay for a 600-μm line versus the via location l_1 for various values of r_{21}. The interconnect parameters are $r_1 = 79.5$ Ω/mm, $r_{v1} = 5.7$ Ω/mm, $c_{v1} = 6$ pF/mm, $c_2 = 439$ fF/mm, $c_{12} = 0.46$, $l_v = 20$ μm, and $n = 2$. The driver resistance and load capacitance are $R_S = 50$ Ω and $C_L = 50$ fF, respectively.

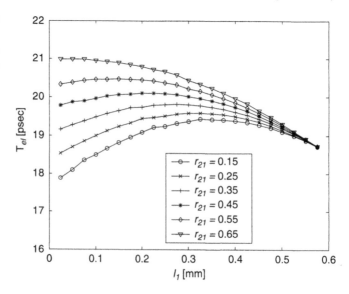

for a 600-μm line are shown as a function of the via location l_1. Note that the delay reaches a maximum and that the maximum shifts to the right as r_{21} increases. If $l_1 < l_d$, the via should be placed closest to the receiver, while if $l_1 > l_r$, the via should be placed closest to the driver.

C) $\frac{d^2 T_{el}}{dl_1^2} = 0$. If the second derivative equals zero, (7-4) becomes

$$T_{el}(l_1) = A_2 l_1 + A_3, \qquad (7\text{-}14)$$

which is a linear function of l_1. The first derivative of (7-4) is equal to A_2. For $A_2 < 0$, (7-4) is strictly decreasing, and the delay is minimum by placing the via closest to the receiver. For $A_2 > 0$, (7-4) is strictly increasing, and the delay is minimum by placing the via closest to the driver. Note that the fundamental minimum distance of a via from a cell is technology dependent. In the special case where $r_{21} = c_{12} = 1$, from (7-6) and (7-7), $A_2 << A_3$ and the delay is independent of l_1. However, as n increases, A_2 also increases, and the choice of n affects the rate of change in the delay. This particular case corresponds to the model introduced in [130] to evaluate the performance of 3-D ICs, which does not capture the timing behavior of interplane interconnects in 3-D circuits as analyzed in cases A and B.

As illustrated in Figures 7-8 and 7-9, the optimum via location shifts to the right (left) when r_{21} increases (decreases). To explain this behavior, consider the definitions of r_{21} and c_{12} in (7-9) and (7-10), where r_{21} (c_{12}) describes the resistance (capacitance) ratio of the corresponding horizontal segments. Referring to Figure 7-8, both r_{21} and c_{12} are greater than one, which means that segment 1 is less (more) resistive (capacitive) than segment 2. Assuming that $r_{21} = 1$, the delay of the line decreases as the length of the more capacitive segment l_1 (i.e., C_1) decreases. However, l_1 does not vanish because C_2 increases as l_1 decreases, approaching C_1. Consequently, as l_1 is decreased beyond a certain distance, described in (7-13), the delay starts to increase. As r_{21} becomes greater than one, the optimum point occurs with increasing values of l_1, although segment 1 is more capacitive than segment 2 ($c_{12} > 1$). This behavior occurs because the delay depends not only on the capacitance, but also on the current, which is controlled by the resistance of each segment. Due to the distributed nature of the impedance of the line, an increase in the resistance of the second segment (increasing r_{21}) can be compensated, such that the delay remains minimum, by reducing the capacitance that this resistance sees near the receiver (the upstream resistance). In the case where $c_{12} >> 1$ and $r_{12} << 1$, and where segment 1 is both more capacitive and resistive than segment 2, l_1

becomes small to reduce the overall delay. In this case, (7-13) yields negative values for l_1 and is essentially zero (the via should be located next to the driver).

7.2.4 **Improvement in Interconnect Delay**

SPICE simulations of the 50% propagation delay of two-terminal nets with a single via are listed in Table 7-1 for various impedance characteristics. The interconnect propagation delay listed in columns 8, 9, and 11 corresponds to the cases where the via is optimally placed; the via is placed at the center of the interconnect and at the maximum delay point. In column 10, the improvement in delay over the case where the via is placed at the center of the line is reported, while the improvement in delay over placing the via at the end points of the line is listed in column 12. If the delay exhibits a minimum according to (7-11), the maximum delay coincides with one of the ends of the interconnect; otherwise, the maximum delay is described by (7-13). The delay analysis considers relatively short interconnect lengths where repeaters cannot reduce the interconnect delay, and therefore optimally placing the vias is the primary way to decrease the delay of these interconnects.

Note that the improvement in delay achieved by optimally placing the via is considerable for those cases where the optimum location coincides with the ends of the interconnect. This large improvement occurs because in these cases the minimum delay, as described by (7-13), occurs at values between the ends of the interconnect, that is, $[0, L]$. For these instances, the propagation delay has an almost linear dependence (with large slope) on the via location, and, therefore, the improvement in delay is significant. Alternatively, for those instances where the optimum occurs for positive values smaller than the interconnect length L, the variation in the delay improvement is small as the delay exhibits a parabolic form. Consequently, the rate of change in the delay with via location is small.

For several interconnect instances, the optimum location is close to either the receiver or the driver. Routing congestion therefore causes some of the interplane vias to be placed at nonoptimal locations. In addition, the improvement in delay that can be achieved varies considerably with the interconnect parameters. In Figures 7–10 and 7–11, the decrease in the improvement in delay resulting from the

Table 7-1 SPICE simulaion results for two-terminal interconnects with a single-interplane via.

Length [μm]	R_S [Ω]	C_L [fF]	r_1 [Ω/mm]	c_2 [fF/mm]	r_{21}	c_{12}	T_{opt}	T_{center}	Impr [%]	T_{max}	Impr [%]
200	100	50	159.5	239	3.50	1.55	19.79	20.02	1.16	20.76	4.90
200	100	50	159.5	239	2.85	1.85	19.89	20.08	0.96	20.44	2.77
300	120	100	93.8	387	2.88	1.67	36.03	36.30	0.75	37.62	4.41
300	120	100	93.8	387	2.88	1.35	34.56	34.80	0.69	35.92	3.94
400	100	100	75.8	287	2.15	2.87	29.30	34.98	19.39	42.34	44.51
400	100	100	75.8	287	3.70	0.77	24.74	28.94	16.98	33.34	34.76
500	30	100	23.8	287	1.50	2.57	54.02	63.37	17.31	72.58	34.36
500	30	100	23.8	287	3.35	2.37	58.18	58.69	0.88	59.77	2.73
500	30	100	23.8	287	3.15	0.87	51.09	54.83	7.32	59.02	15.52
Average Improvement									7.27		16.43

nonoptimal placement of the vias is illustrated for a 500-μm inter-plane interconnect. The ratios r_{21} and c_{12} range from 0.1 to 5, and the interplane via is placed 50 μm from the optimum location.

A nonnegligible reduction in the improvement in delay can occur from the nonoptimal placement of the via for a 50-μm departure from the optimal location. In addition, the decrease in improvement is greater for those values of r_{21} and c_{12}, where the optimum is at the ends of the line. For the optimum location corresponding to points along the interconnect length, the decrease in the delay improvement is small as depicted by the approximately flat portion of the surface shown in Figures 7-10 and 7-11. In addition, the dependence of the delay improvement on the capacitance ratio c_{12} is in general greater than the dependence on the resistance ratio r_{12}. For low values of the driver resistance, the dependence of the delay improvement on r_{12} is significant, as shown in Figure 7-10. Increasing the driver resis-tance lowers this dependence, as depicted in Figure 7-11, as the term $R_S \sum_i C_i$ dominates the delay of the line.

This dependence shows that for large drivers, even if the interconnect is almost uniform ($r_{21} \approx 1$), considerable savings in delay can be achieved by optimally placing the via. Furthermore, as larger drivers are used to reduce the interconnect delay, the improvement in delay becomes greater. If small drivers are utilized, the savings in delay is obtained primarily from the difference in the capacitance of the seg-ments, even if $c_{12} \approx 1$. In addition, a slight decrease in the maximum

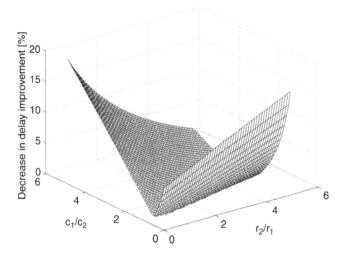

■ **FIGURE 7-10** Decrease in the delay improvement caused by the nonoptimal placement of the interplane via for a 500-μm interconnect. The interconnect parameters are $r_1 = 23.5$ Ω/mm, $r_{v1} = 270$ Ω/mm, $c_{v1} = 270$ fF/mm, $c_2 = 287$ fF/mm, $l_v = 15$ μm, and $n = 2$. The driver resistance and load capacitance are $R_S = 30$ Ω and $C_l = 100$ fF, respectively.

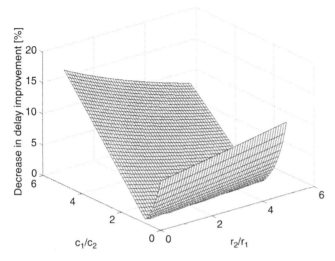

■ **FIGURE 7-11** Decrease in the delay improvement due to the nonoptimal placement of the interplane via for a 500-μm interconnect. The interconnect parameters are $r_1 = 23.5 \, \Omega/\text{mm}$, $r_{v1} = 6.7 \, \Omega/\text{mm}$, $c_{v1} = 270 \, \text{fF/mm}$, $c_2 = 287 \, \text{fF/mm}$, $l_v = 15 \, \mu\text{m}$, and $n = 2$. The driver resistance and load capacitance are $R_S = 100 \, \Omega$ and $C_L = 100 \, \text{fF}$, respectively.

degradation of the improvement in delay that occurs from a non-optimal via placement can be observed in Figures 7-9 and 7-10. This decrease occurs due to the increase in the driver resistance, which is considered fixed and can greatly affect the interconnect delay, while the delay variation due to the impedance characteristics of the segments is reduced. A similar behavior applies for the capacitance ratio c_{12} where the load capacitance is increased. In the following section, two-terminal nets with multiple-interplane vias are investigated.

7.3 TWO-TERMINAL INTERCONNECTS WITH MULTIPLE-INTERPLANE VIAS

In the previous section, the case where a two-terminal interconnect includes only one interplane via is analyzed. Due to routing obstacles and placed cells, however, a stacked interplane via that spans multiple physical planes may not be possible. In addition, in the previous analysis, no constraints are imposed on the location of the via. This situation is rather impractical since existing interconnects and active devices may not permit an interplane via to be placed at a particular location along the interconnect line. In addition, since these vias penetrate the device layers, certain regions are reserved for interplane via placement to minimally disrupt the location of the transistors.

■ **FIGURE 7-12** Interplane interconnect consisting of m segments connecting two circuits located n planes apart.

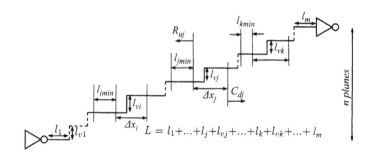

$$L = l_1 + \ldots + l_j + l_{vj} + \ldots + l_k + l_{vk} + \ldots + l_m$$

Due to these reasons, a two terminal interplane interconnect in a 3-D circuit consists of multiple-interplane vias where each one of these vias is placed within a certain physical interval. A schematic of an interplane interconnect connecting two circuits located n planes apart is illustrated in Figure 7-12. The horizontal segments of the line are connected through the vias, which can traverse more than one plane. Consequently, the number of horizontal segments within the interconnect m is smaller or at most equal to the number of physical planes between the two circuits, that is, $n \geq m$, where the equality only applies when each of the vias connects metal layers from two adjacent physical planes.

Each horizontal segment j of a line is located on a different physical plane with length l_j. The vias are denoted by the index of the first of the two connected segments. For example, if a via connects segment j and $j + 1$, the via is denoted as v_j with length l_{vj}. Note that planes j and $j + 1$ are not necessarily physically adjacent. The total length of the line L is equal to the summation of the length of the horizontal segments and vias,

$$L = l_1 + l_{v1} + \ldots + l_j + l_{vj} + \ldots + l_n. \tag{7-15}$$

The length of each horizontal segment of a line is bounded,

$$l_{j\,min} \leq l_j \leq l_{j\,min} + \Delta x_j, \text{ for } j = 1, \tag{7-16a}$$

$$l_{j\,min} \leq l_j \leq \Delta x_{j-1} + l_{j\,min} + \Delta x_j \text{ for } j \in [2, m-1], \tag{7-16b}$$

$$l_{j\,min} \leq l_j \leq l_{j\,min} + \Delta x_{j-1}, \text{ for } j = m, \tag{7-16c}$$

and the via placement is also constrained,

$$0 \leq x_j \leq \Delta x_j, \tag{7-17}$$

where l_{jmin} denotes the minimum length of the interconnect segment on plane j, and Δx_j is the length of the interval where the via that connects planes j and $j + 1$ is placed. This interval length is called the "allowed interval" here for clarity. x_j is the distance of the via location from the edge of the allowed interval.

l_{jmin} is the length of an interconnect segment connecting two allowed intervals or an allowed interval and a placed cell. These lengths are considered fixed. Alternatively, the routing path of a net is not altered except for the via location within the allowed interval. Each horizontal segment is assumed to be on a single metal layer within the physical plane. In the case where a horizontal segment is on more than one layer, as the outcome of a layer assignment algorithm [230], the problem can be approached in two different ways. The intraplane vias can be treated as additional variables where the location of these vias needs to be determined. This formulation requires, however, the generation of additional allowed intervals, specifically for the intraplane vias. Alternatively, the first and last sections of the segment connected to the interplane vias remain as a variable while the remaining sections of the horizontal segment constitute the minimum length of segment l_{jmin}, which is constant.

The distributed Elmore delay model is used to determine the delay of these interconnects, and the corresponding electrical model of the line is depicted in Figure 7-13. The related notation is listed in Table 7-2. The distributed Elmore delay of a two-terminal interconnect in matrix form is

$$T(\mathbf{l}) = 0.5 \ \mathbf{l}^T \mathbf{A} \mathbf{l} + \mathbf{b} \mathbf{l} + D, \tag{7-18}$$

$$\mathbf{l} = \begin{bmatrix} l_1 & l_2 & \cdots & l_{m-1} & l_m \end{bmatrix}^T, \tag{7-19}$$

$$\mathbf{A} = \begin{bmatrix} r_1 c_1 & r_1 c_2 & \cdots & r_1 c_m \\ \vdots & \vdots & \cdots & \vdots \\ r_1 c_m & r_2 c_m & \cdots & r_m c_m \end{bmatrix}, \tag{7-20}$$

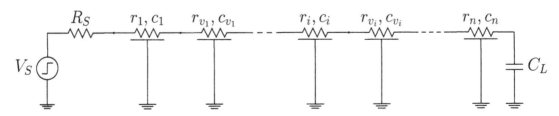

■ FIGURE 7-13 Interplane interconnect model composed of a set of nonuniform distributed *RC* segments.

Table 7-2 Notation for two-terminal interplane interconnects.

Notation	Definition
R_S	Driver resistance
C_L	Load capacitance
r_j (c_j)	Resistance (capacitance) per unit length of horizontal segment j
r_{vj} (c_{vj})	Resistance (capacitance) per unit length of interplane via v_j
R_j (C_j)	Total interconnect resistance (capacitance) of horizontal segment j
R_{vj} (C_{vj})	Total interconnect resistance (capacitance) of interplane via v_j
R_{u_j}	Upstream resistance of the allowed interval for via v_j
C_{d_j}	Downstream capacitance of the allowed interval for via v_j

$$\mathbf{b} = \begin{bmatrix} r_1 \left(\sum_{i=1}^{m-1} c_{vi} l_{vi} + C_L \right) + c_1 R_S \\ \vdots \\ r_m C_L + c_m \left(R_S + \sum_{i=1}^{m-1} r_{vi} l_{vi} \right) \end{bmatrix}^{\mathrm{T}}, \tag{7-21}$$

$$D = R_S \sum_{i=1}^{m-1} c_{vi} l_{vi} + C_L \sum_{i=1}^{m-1} r_{vi} l_{vi} + \frac{1}{2} \sum_{i=1}^{m-1} r_{vi} c_{vi} l^2 + R_S C_L. \tag{7-22}$$

Since (7-9) is a constant quantity, the optimization problem can be described as follows,

(P)   minimize $T(\mathbf{l}) = 0.5 \, \mathbf{l}^{\mathrm{T}} \mathbf{A} \mathbf{l} + \mathbf{b} \mathbf{l}$,
      subject to (7-15) - (7-17).

As described by the following theorem, the primal problem (P) is typically not convex, and therefore convex quadratic programming optimization techniques are not directly applicable.

Theorem 1: The primal optimization problem (P) is convex *iff*

$$r_{j+1} c_j - r_j c_{j+1} > 0. \tag{7-23}$$

Proof: $\mathbf{A}$ is a positive definite matrix if all subdeterminants are positive. By elementary row operations, the subdeterminants of $\mathbf{A}$ are positive *iff* (7-23) applies. If (7-23) applies, $\mathbf{A}$ is positive definite, and (P) is a convex optimization problem.

Note that condition (7-23) must be satisfied for every segment j such that (P) is a convex optimization problem. Due to the heterogeneity of 3-D circuits, (7-18) typically is of indefinite quadratic form. Therefore, convex quadratic programming techniques may not produce the global minima. Certain transformations can be applied to convert (P) into a convex optimization problem [231]; the objective functions, however, are no longer quadratic. Alternatively, (P) can be treated as a geometric programming problem. Geometric programs include optimization problems for functions and inequalities of the following form,

$$g(y) = \sum_{j=1}^{M} s_j y_1^{a_{1j}} y_2^{a_{2j}} \cdots y_n^{a_{nj}}, \qquad (7\text{-}24)$$

$$s_j y_1^{a_{1j}} y_2^{a_{2j}} \cdots y_n^{a_{nj}} \leq 1, \qquad (7\text{-}25)$$

where the variables y_j's and coefficients s_j's must be positive and the exponents a_{ij}'s are real numbers. Although equality constraints are not allowed in standard geometric problems, (P) can be solved as a generalized geometric program as described in [232] where a globally optimum solution is determined. Alternatively, by considering the particular characteristics of the optimization problem, an efficient heuristic is proposed for placing the interplane vias.

7.3.1 Two-Terminal Via Placement Heuristic

In this section, a heuristic for the near-optimal interplane via placement of two-terminal nets that include several interplane vias is described. The key concept in the heuristic is that the optimum via placement depends primarily on the size of the allowed interval (that is estimated or known after an initial placement) rather than the exact location of the via. Consider the interplane interconnect line shown in Figure 7-12, where the optimum location for via j connecting interconnect segments j and $j + 1$ is to be determined. With respect to this via, the critical point (*i.e.*, $\frac{\partial T_{el}}{\partial x_j} = 0$) of the Elmore delay is

$$x_j = - \left[\frac{l_{vj}\left(r_j c_{vj} - r_{vj}c_{j+1} + r_{j+1}c_{j+1} - r_j c_{j+1}\right) + R_{uj}\left(c_j - c_{j+1}\right) + \Delta x_j\left(r_j - r_{j+1}\right)c_{j+1} + C_{di}\left(r_j - r_{j+1}\right)}{r_j c_j - 2r_j c_{j+1} + r_{j+1}c_{j+1}} \right],$$

$$(7\text{-}26)$$

where R_{uj} and C_{dj} are the upstream resistance and downstream capacitance, respectively, of the allowed interval for via j (see also Table 7-2), as shown in Figure 7-12. The Elmore delay of the line with respect to x_j can be either a convex or a concave function. The remaining discussion in this section applies to the case where the Elmore delay of the line is a convex function with respect to x_j. A similar analysis can be applied for the concave case.

In (7-26), the optimum via location x_j^* is a monotonic function of R_{uj} and C_{dj}. The sign of the monotonicity depends on the interconnect impedance parameters of the segments j and $j+1$ connected by via j. As the size of the allowed intervals for all of the vias is constrained by (7-17), the minimum and maximum value of R_{uj} and C_{dj} can be readily determined, permitting the values of x_j^* for these extrema, $x_{j\min}^*$ and $x_{j\max}^*$, to be evaluated. Due to the monotonic dependence of x_j on R_{uj} and C_{dj}, the optimum location for via j, x_j^*, lies within the range delimited by $x_{j\min}^*$ and $x_{j\max}^*$. Four cases can be distinguished:

1. If $x_{j\max}^* \leq 0$, $x_j^* = 0$, and the optimum via location coincides with the lower bound of the interval as defined by (7-17).
2. If $x_{j\min}^* \geq \Delta x_j$, $x_j^* = \Delta x_j$, and the optimum via location coincides with the upper bound of the interval as defined by (7-17).
3. If $\Delta x_j \geq x_{j\min}^* \geq 0$ and $\Delta x_j \geq x_{j\max}^* \geq 0$, the bounded interval as defined by (7-17) reduces to

$$0 \leq x_{j\min}^* \leq x_j \leq x_{j\max}^* \leq \Delta x_j. \tag{7-27}$$

In this case, the via location cannot be directly determined. However, by iteratively decreasing the range of values for x_j^*, the optimal location for via j can be determined.

The following example is used to demonstrate that the physical domain for x_j^* iteratively decreases to a single point, the optimum via location. A detailed analysis of the heuristic is provided in Appendix C. Consider segments i, j, and k depicted in Figure 7-12, where segments i and k are located upstream and downstream of segment j, respectively. From (7-16) and (7-17), the minimum and maximum values of R_{ui}^0, R_{uj}^0, R_{uk}^0, C_{di}^0, C_{dj}^0, and C_{dk}^0 are determined, where the superscript represents the number of iterations. Assume that $x_{\min}^{*0}$ and $x_{\max}^{*0}$ are obtained from (7-26) to satisfy (7-27) for all three segments i, j, and k. As the range of values for the via location of segments i and k decreases according to (7-27), the minimum (maximum) value of the upstream resistance and downstream capacitance of segment j increases (decreases), that is, $R_{uj\min}^0 < R_{uj\min}^1$, $C_{dj\min}^0 < C_{dj\min}^1$, $R_{uj\max}^1 < R_{uj\max}^0$, and $C_{dj\max}^1 < C_{dj\max}^0$. Due to the monotonicity of x_j^* on R_{uj} and C_{dj}, $x_{j\min}^{*0} < x_{j\min}^{*1}$ and $x_{j\max}^{*1} < x_{j\max}^{*0}$. The range of values for x_j^* therefore also decreases, and, typically, after two or three iterations, the optimum location for the corresponding via is determined. In Figure 7-14, the convergence of the

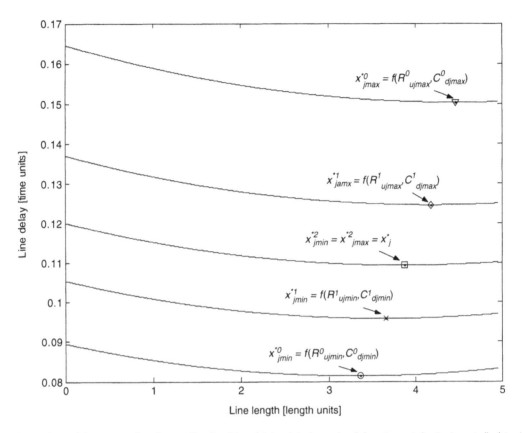

The y-axis is labeled "Line delay [time units]" and the x-axis is labeled "Line length [length units]".

Labels in the figure:
$x^{*0}_{jmax} = f(R^0_{ujmax}, C^0_{djmax})$

$x^{*1}_{jamx} = f(R^1_{ujmax}, C^1_{djmax})$

$x^{*2}_{jmin} = x^{*2}_{jmax} = x^*_j$

$x^{*1}_{jmin} = f(R^1_{ujmin}, C^1_{djmin})$

$x^{*0}_{jmin} = f(R^0_{ujmin}, C^0_{djmin})$

■ **FIGURE 7-14** Case 3 of the two-terminal net heuristic. The allowed interval is iteratively decreased until the optimum via location is eventually determined.

heuristic at the optimum via location is depicted for one via of a two-terminal interconnect. For this example, $c_j > c_{j+1}$ and $r_j > r_{j+1}$, and, therefore, $x^*_{j\,min} = f(R_{j\,min}, C_{j\,min})$ and $x^*_{jmax} = f(R_{jmax}, C_{jmax})$. As shown in Figure 7-14, the heuristic converges to the optimum via location within several iterations.

4. if $x^*_{j\,min} \leq 0$ and $x^*_{j\,min} \geq \Delta xj$, the via location cannot be directly determined. In addition, the bounding interval cannot be reduced. Consequently, some loss of optimality occurs. This departure from the optimal, however, is typically smaller than a few percent for all of the tested conditions, as shown by the results presented in subsection 7.3.3.

The variation between the extrema of the upstream resistance and downstream capacitance due to the relatively small size of the allowed interval of the via placement ensures that a significant

■ **FIGURE 7-15** A subset of interconnect instances depicted by the dashed lines for case 4 of the via placement heuristic. The interconnect traverses eight planes and has a length $L = 1.455$ mm. The resistance r_j and capacitance c_j of each interconnect segment range from 10 Ω/mm to 50 Ω/mm and 100 fF/mm to 500 fF/mm, respectively.

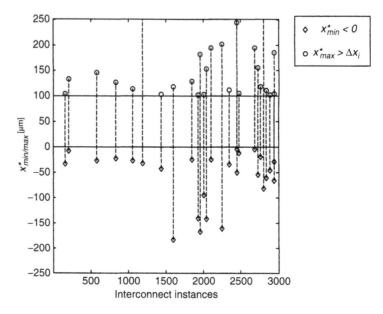

variation between the values $x_{j\min}^*$ and $x_{j\max}^*$ does not occur for most interconnects. To justify this argument, the extreme values of x_j^* are plotted in Figure 7-15 for the fourth interplane via of a two-terminal net considered to traverse eight planes. This example represents a significant variation in the extreme values of x_j^*, due to the location of the via and the large number of vias along the interconnect. Indeed, the large number of vias (seven in this example) implies a considerable variation in the value of the downstream capacitance and upstream resistance, particularly for the fourth via. Note that the last via of the interconnect, for instance, only sees a variation in the upstream resistance. As shown in Figure 7-15, less than 1% of the interconnect instances yield such boundary values for x_j^*, where the inequalities $x_{j\min}^* \leq 0$ and $x_{j\max}^* \geq \Delta x_j$ are satisfied. This small percent (1% in this case) is similar or smaller for the other vias of the interconnect due to the smaller variation in the upstream resistance or downstream capacitance. In addition, the inequalities, $x_{j\min}^* \leq 0$ and $x_{j\max}^* \geq \Delta x_j$, are typically satisfied, where the size of the allowed interval for a via is relatively small as compared to the size of the allowed intervals for the remaining vias. A nonoptimal via placement for that interconnect segment does not significantly affect the overall delay of the line.

Furthermore, for case 4 to occur, there should be at least two vias for which an optimum location cannot be found. Indeed, if all but one of the vias are placed, the exact value of the upstream (downstream) resistance (capacitance) for the nonoptimized via can be obtained and, therefore, the location of that via can be determined from (7-26). The nonoptimal placement of a via does not necessarily affect the optimal placement of the remaining vias. For example, any via placed according to the criteria described in cases 1 and 2 is not affected by the placement of the remaining vias. Therefore, as noted earlier, the size of the allowed intervals rather than the exact location of the vias is the key factor in determining the optimum via locations. Based on this heuristic, an algorithm for placing a via in two-terminal nets is presented in the following subsection, along with simulation and analytic results for these interconnects.

7.3.2 Two-Terminal Via Placement Algorithm

The heuristic described in the previous subsection has been used to implement an algorithm that exhibits an optimal or near-optimal via placement for two-terminal interplane interconnects in 3-D ICs, with significantly lower computational time as compared to geometric programming solvers. The pseudocode of the two-terminal via placement algorithm (TTVPA) is illustrated in Figure 7-16. The input to the algorithm is a description of the interplane interconnect where the minimum length of the segments and the size of the allowed intervals are provided. In the first step of the algorithm, the array of the maximum and minimum upstream (downstream) resistance (capacitance) for

Two-Terminal Via Placement Algorithm: (l_{min}, Δx)
1. Determine C_{dmin}, C_{dmax}, R_{umin}, R_{umax},
2. while $S \neq \emptyset$
3.   if iter < max_iter
4. $s_j \leftarrow$ an uprocessed via
5.   obtain $x^*_{j\,min}$ and $x^*_{j\,max}$ from eq. (7.26)
6.   check for the inequalities in (*i*)–(*iv*)
7.   if s_j is optimized (*cases i-ii*)
8.   store optimum via location
9. $s \leftarrow s - \{s_j\}$
10.   update C_{dmin}, C_{dmax}, R_{umin}, R_{umax},
elseif Δx_j decreases (*case iii*)
11.   update l_{minj}, C_{dmin}, C_{dmax}, R_{umin}, R_{umax},
else (case iv)
12.   go to step 3
else (*the non-optimized vias*)
13.   place via at the center of the allowed interval
14.   store via location
15. $s \leftarrow s - \{s_j\}$
16. exit

■ **FIGURE 7-16** Pseudocode of the proposed Two-Terminal Net Via Placement Algorithm (TTVPA).

every allowed interval is determined. In the following steps, for each unprocessed via, the range of values for the optimum via location as given by (7-26) is evaluated. In step 5, these values are compared to the inequalities described in the previous subsection. If an optimal via location is determined in this step, the via is marked as processed and the capacitance and resistance arrays are updated. If, after a number of iterations, there are nonoptimal vias, in step 14 these vias are placed at the center of the corresponding allowed intervals and the algorithm terminates. Other criteria, such as routing congestion, can alternatively be applied to place nonoptimal vias rather than placing these vias at the center of the corresponding allowed intervals. Further criteria can be considered to search for the optimum location of the nonoptimal vias, trading off runtime with accuracy.

7.3.3 **Application of the Via Placement Technique**

The via placement algorithm has been applied to several interconnect instances to validate the accuracy and efficiency of the heuristic. Both SPICE simulations and optimization results are provided. Two-terminal interplane interconnects for different numbers of physical planes are analyzed. The impedance characteristics of the horizontal segments and vias are extracted for several interconnect structures using a commercial impedance extraction tool [224]. Copper interconnect has been assumed with an effective resistivity of 2.2 $\mu\Omega$-cm. Based on the extracted impedances, the resistance and capacitance of the horizontal segments range from 25 Ω/mm to 125 Ω/mm and 100 fF/mm to 300 fF/mm, respectively, for a 90-nm technology node [233], [234]. The cross section of the vias is 1 μm $\times$ 1 μm, with 1-μm spacing from the surrounding horizontal metal layers assuming a silicon on insulator (SOI) process, as described in [140]. For all of the interconnect structures, the total and minimum length of each horizontal segment is randomly generated. For simplicity, all of the vias connect the segments of two adjacent physical planes (*i.e.*, $m = n$).

Delay simulations are reported in Table 7-3. The delay of the line T_{center}, where the vias are placed at the center of the allowed intervals, is listed in column 2. The delay T_{rnd}, listed in column 3, corresponds to the line delay for random via placement. The minimum interconnect delay T_{min}, where the vias are optimally placed, is listed in column 4. The via locations or, equivalently, the length of the horizontal segments, are determined from the via placement algorithm. The improvement in delay as compared to the case where the vias are placed at the center of the line is listed in column 5. The number

Table 7-3 SPICE simulation results demonstrating the delay savings achieved by near optimal via placement. The resistance and capacitance per unit length of the vias are $r_{vi} = 6.7 \ \Omega/\text{mm}$ and $c_{vi} = 6 \ \text{pF/mm}$, respectively. The length of the vias is $L_{vi} = 20 \ \mu\text{m}$. The driver resistance is $R_s = 15 \ \Omega$, and the load capacitance is $C_L = 100 \ \text{fF}$. The length of the allowed intervals is $\Delta x_i = 200 \ \mu\text{m}$.

Length [μm]	T_{center} [ps]	T_{rnd} [ps]	T_{min} [ps]	Improvement [%]	n
1017	12.35	12.64	11.42	8.14 (10.68)	4
1180	13.37	14.42	12.33	8.43 (16.95)	4
849	11.00	11.71	10.27	7.11 (14.02)	4
969	13.52	14.96	12.12	11.55 (23.43)	4
967	12.38	12.59	11.72	5.63 (7.42)	4
1612	18.54	19.85	17.24	7.54 (15.14)	5
1537	20.80	19.47	19.37	7.38 (0.52)	5
1289	17.78	18.43	16.45	8.09 (12.04)	5
1443	18.77	19.54	18.07	3.87 (8.14)	5
1225	16.97	18.33	15.62	8.64 (17.35)	5
2118	30.52	34.81	26.44	15.43 (31.66)	7
2130	27.92	27.32	25.94	7.63 (5.32)	7
1961	28.49	30.67	26.16	8.91 (17.24)	7
2263	35.58	40.11	31.31	13.64 (28.11)	7
2174	32.31	30.34	29.16	10.80 (4.05)	7
Average Improvement				8.85 (14.14)	

in parentheses corresponds to the improvement in delay over a random via placement. Note that the variation in the improvement in delay changes significantly for those listed instances, although the interconnect lengths are similar and the load capacitance and driver resistance are the same. This considerable variation demonstrates the strong dependence of the line delay on the impedance characteristics of the segments of the line and supports modeling the interplane interconnect as a group of nonuniform segments. In addition, depending on the impedance characteristics of the line segments, placing a via at the center of the allowed intervals is, for certain instances, near-optimal, explaining why the improvement in delay is not significant in these instances. The same characteristic applies to those cases where a random placement is close to the optimum placement. Nevertheless, as listed in Table 7-3, an improvement of up to 32% is observed for relatively short interconnects, demonstrating that an optimum via placement can significantly enhance the speed of 3-D circuits (in addition to the primary benefit of reduced wirelength and therefore lower power).

The two-terminal via placement algorithm is compared in terms of both optimality and efficiency to two optimization solvers. The first solver, YALMIP [235], is a general optimization solver that supports geometric programming, while GLOPTIPOLY [236] is an optimization solver for nonconvex polynomial functions. Due to the excessive computational time of GLOPTIPOLY (greater than three orders of magnitude as compared to YALMIP), only comparisons with YALMIP are reported. Optimization results are listed in Table 7-4 for different values of Δx ranging from 50 µm to 300 µm.

As reported in column 9 of Table 7-4, TTVPA exhibits high accuracy as compared to YALMIP. These results are independent of the number of planes that comprise the 3-D interconnect, demonstrating that TTVPA yields optimum solutions for most interconnect instances. In addition, for those cases where some of the vias are not optimally placed, the loss of optimality is insignificant (as previously discussed). In column 8, the runtime ratio of YALMIP to TTVPA is listed. TTVPA is approximately two orders of magnitude faster than YALMIP. The complexity of TTVPA has an almost linear dependence on the number of interplane vias. As depicted in Figure 7-16, each via is typically processed once; otherwise, a maximum of two to five iterations is required to place a via. In all of the experiments, the algorithm terminates within a number of iterations smaller than five times the total number of vias within the interconnect.

As shown in Table 7-4, the savings in delay from the near-optimal via placement strongly depends on the length of the allowed intervals. For example, doubling the length of the allowed intervals for via placement increases almost twofold the maximum improvement in delay. As the length of the allowed intervals increases, the constraints in (7-16) are relaxed, and a greater benefit from optimally placing the vias is achieved.

The effect of the nonuniformity of the interplane interconnects on the improvement in delay is graphically illustrated in Figure 7-17, where the improvement in delay for interplane interconnects spanning four and five physical planes is depicted. The average savings

Table 7-4 Optimization results for various two-terminal interplane interconnects and numbers of physical planes n.

n	Average Interconnect Length [μm]	Δx_i's [μm]	Delay Improvement [%] Vias Placed at the Center		Random Via Placement		YALMIP/ TTVPA Runtime ratio × times	Max. error [%]	Instances
			Avg	Max	Avg	Max			
3	270	50	3.36	11.10	5.88	22.24	141.7	0.005	10000
3	520	100	4.59	17.63	5.02	17.63	148.6	0.008	10000
3	1020	200	5.90	23.12	10.27	47.08	145.6	0.013	10000
4	405	50	4.02	13.01	6.00	25.97	209.4	0.006	10000
4	781	100	5.26	16.95	7.91	34.10	574.3	0.003	10000
4	996	100	3.45	9.87	5.37	20.05	95.5	0.008	5000
4	1155	150	5.94	21.61	8.89	44.46	125.4	0.011	10000
4	1302	200	6.31	22.76	9.87	46.91	241.6	0.021	5000
4	1600	300	8.73	26.85	12.58	55.91	111.4	0.025	5000
5	540	50	4.48	13.73	6.16	27.49	112.14	0.005	10000
5	1040	100	5.69	17.97	7.79	35.82	335.40	0.012	10000
5	1541	150	6.35	22.36	8.63	46.26	549.35	0.017	10000
5	1277	100	3.91	10.84	5.57	22.04	93.1	0.003	5000
5	1684	200	7.06	19.65	10.09	41.28	226.4	0.009	5000
5	2076	300	9.77	26.74	14.27	55.45	90.1	0.009	5000
7	1840	100	4.64	12.81	6.21	25.16	84.6	0.008	5000
7	2440	200	8.26	23.77	11.07	47.56	89.0	0.002	5000

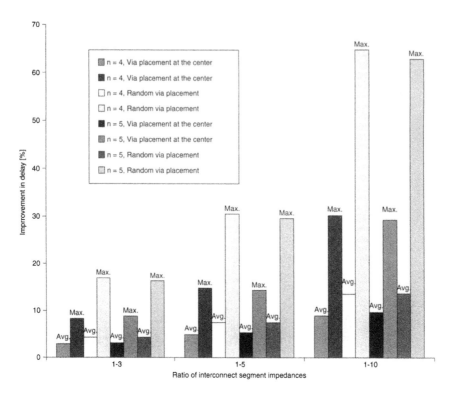

■ **FIGURE 7-17** Average and maximum improvement in delay for different ranges of interconnect segment resistance and capacitance ratios. (a) The vias are placed at the center of the allowed intervals and (b) the vias are randomly placed.

in delay of highly non-uniform interconnects (*i.e.*, $r_{(j+1)} / r_j = 1 - 10$ and $c_j / c_{(j+1)} = 1 - 10$) can be significant, approaching 10% and 13% for a moderately sized length, where the vias are placed at the center of the allowed intervals and are randomly placed, respectively. The maximum improvement can exceed 60%, as shown in Figure 7-17.

The length of most of the interconnects listed in Tables 7-1 to 7-4 is such that repeater insertion cannot improve the performance of these interconnects. Alternatively, as mentioned earlier, wire sizing or a via placement technique can be utilized to improve the speed of these interconnects. Wire sizing is a well-known technique used to reduce interconnect delay. Wire shaping, however, is not always feasible due to routing congestion or obstacles such as placed cells. In addition, as the interconnect is widened to lower the interconnect resistance, the capacitance and, consequently, the power consumption of the interconnect increase.

The wire sizing algorithm described in [237] has been applied to several interconnects to improve the line delay. The interconnect length is divided equally among the horizontal segments that constitute the interconnect. For the same interconnects, the line delay where the width is minimum and the vias are optimally placed is also determined. In Figure 7-18, the average interconnect delay for both the via placement and wire sizing technique is shown. The instance where the optimum via placement outperforms wire sizing (and vice versa) is depicted. The average delay improvement ranges from 6.23% for $n = 4$ to 17.8% for $n = 5$, justifying that via placement can be an effective delay reduction technique for interplane interconnects in 3-D circuits, without requiring additional area. Wiring sizing does not achieve a significant delay reduction primarily because of the via impedance characteristics and because the vias cannot be sized as aggressively as the horizontal segments. Furthermore, via sizing is not particularly desirable as wider vias decrease the via density or, equivalently, the number of interplane interconnects that can be routed throughout a 3-D system.

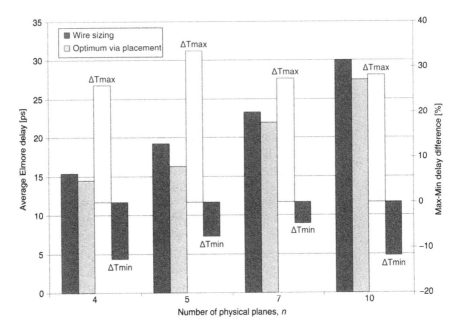

■ **FIGURE 7-18** Comparison of the average Elmore delay based on wire sizing and optimum via placement techniques. The instance where the optimum via placement outperforms wire sizing (and vice versa) is also depicted.

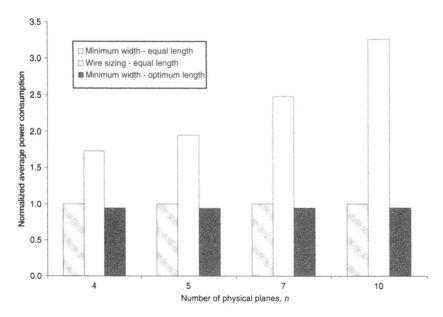

■ **FIGURE 7-19** Normalized average power consumption for minimum width and equal length, wire sizing and equal length, and minimum width and optimum via placement.

The normalized average power consumption (NAPC) for various interconnects is illustrated in Figure 7-19. The crosshatched bar corresponds to minimum-width interconnects and horizontal segments of equal length. The white bar considers those interconnects where the horizontal segments are of equal length and wire sizing has been applied. The NAPC where the line segments are of minimum width and optimum length is depicted by the gray bar. The TTVPA technique dissipates lower power as compared to the other two approaches. Wire sizing increases the capacitance of the line, resulting in an NAPC greater than the optimum via placement technique. Note that, although the NAPC is lowest for optimally placed vias, the capacitance and thus the NAPC are not minimum as the time constant for the interconnect also depends on the resistance of the line. Due to this dependence, the capacitance is not designed to be minimum avoiding an increase in the interconnect resistance. The reduction in NAPC is of considerable importance due to pronounced thermal effects in 3-D circuits, as discussed in Chapter 6.

From the results presented in this section, exploiting the nonuniform impedance characteristics of the interplane interconnects when placing the vias can improve the delay of the lines. This improvement in delay can decrease the number of repeaters required to drive a global line or eliminate repeaters in semi-global lines. In addition, wire sizing can be avoided, thereby saving significant power. Decreasing the number of repeaters and avoiding wide lines reduce the overall power consumption, which is an important advantage in 3-D circuits due to thermal issues.

The overhead caused by placing the vias is the additional effort for placement and routing to generate an allowed interval for the vias, which increases the routing congestion. Other techniques, however, also require similar if not greater overhead. Repeaters, for example, consume silicon area, dissipate power, and block the metal layers within a physical plane when driving nets on the topmost metal layer. Wire sizing requires routing resources reserved for wider interconnect segments. A discussion on placing vias for the important class of multiterminal nets is deferred to the following chapter.

7.4 **SUMMARY**

A technique for timing-driven placement of interplane vias in two-terminal 3-D interconnects is presented in this chapter. The key points of this technique can be summarized as follows:

- Interplane interconnect models of 3-D circuits vary considerably from traditional 2-D interconnect models. This deviation is due to several reasons, such as the heterogeneity of 3-D circuits, diverse fabrication technologies, and the variety of bonding styles.

- For an interplane interconnect that includes only one interplane via, a closed-form solution is provided for placing a via that minimizes the Elmore delay.

- For interplane interconnect comprising multiple vias, conditions are provided that minimize the delay of a line by placing interplane vias.

- Geometric programming can be utilized to produce the globally optimum solution of the via placement problem.

- An accurate and efficient heuristic as compared to geometric programming solvers is described. The fundamental concept of the heuristic is that via placement depends primarily on the size of the allowed intervals where the vias can be placed, not on the precise location of the other vias along the line.

- The improvement in delay depends on the size of the allowed intervals and the interconnect impedance characteristics.

- The proposed via placement technique is compared to a wire sizing technique and exhibits a greater average improvement in interconnect performance. Wire sizing cannot effectively handle the nonuniformity of the interplane interconnects and the discontinuities due to the interplane vias.

- The via placement technique decreases the dynamic power consumed by the interconnects as compared to wire sizing techniques, since wider lines are not necessary.

- Timing-driven via placement can be an alternative to repeater insertion to improve the speed of 3-D interconnect systems.

Chapter 8

Timing Optimization for Multiterminal Interconnects

A significant improvement in the performance of two-terminal inter-plane interconnects in 3-D circuits is demonstrated in the previous chapter. A technique that accurately determines the via location that minimizes the delay of a two-terminal interconnect is described and applied to numerous interconnects. Multiterminal interconnects, however, constitute a significant portion of the interconnects in an integrated circuit. Improving the performance of these nets in 3-D circuits is a challenging task as the sinks of these interconnects can be located on different physical planes. In addition to decreasing the delay, a timing optimization technique should not significantly affect the routed tree. A via placement technique for interplane trees is presented in this chapter. The task of placing the vias in an interplane tree to decrease the delay of a tree is described in the following section. A heuristic solution to this problem is described in Section 8.2. Algorithms based on this heuristic are presented in Section 8.3. The application of these algorithms to various interplane trees is discussed in Section 8.4. Finally, a summary is provided in Section 8.5.

8.1 TIMING-DRIVEN VIA PLACEMENT FOR INTERPLANE INTERCONNECT TREES

The problem of placing vias in interplane trees to decrease the delay of these trees is investigated in this section. A simple interplane interconnect tree (also called an interconnect tree for simplicity) is illustrated in Figure 8-1a, while some related terminology is listed in Table 8-1. The sinks of the tree are located on different physical planes within a 3-D stack. Subtrees not directly connected to the interplane vias that do not contain any interplane vias (*i.e.*, intra-plane trees) are also shown. The interconnect segments from each physical plane are denoted by a solid line of varying thickness.

■ **FIGURE 8-1** Interplane interconnect tree: (a) typical interplane interconnect tree and (b) intervals and directions that the interplane via can be placed.

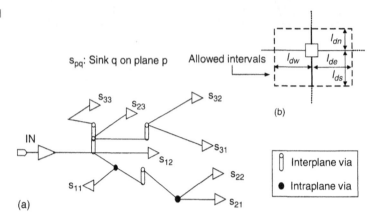

Table 8-1 Notation for two-terminal nets and interconnect trees.

Notation	Definition
R_S	Driver resistance
C_L	Load capacitance
r_j (c_j)	Resistance (capacitance) per unit length of horizontal segment j
r_{vj} (c_{vj})	Resistance (capacitance) per unit length of interplane via v_j
R_j (C_j)	Total interconnect resistance (capacitance) of horizontal segment j
R_{vj} (C_{vj})	Total interconnect resistance (capacitance) of interplane via v_j
R_{u_j}	Upstream resistance of the allowed interval of via v_j
$R_{u_{ij}}$	Common upstream resistance of the allowed interval of via v_i and v_j
d_i	Candidate direction for a type-2 move
C_{d_j}	Total downstream capacitance of the allowed interval of via v_j (in every direction d_i)
$P_{s_{pq}}$	Path from root of the tree to sink s_{pq}
$P_{s_{pq}v_j}$	Path to sink s_{pq} including v_j in every candidate direction
U_{kj}	Set of vias located upstream v_j up to v_k, including v_k and belonging to at least one path $P_{s_{pq}v_j}$
$\overline{P_{s_{pq}v_j}}$	Path to sink s_{pq} that does not include v_j
$\overline{P_{s_{pq}U_{kj}}}$	Path to sink s_{pq} that does not include any of the vias in the set U_{kj}
$P_{s_{pq}v_jd_i}$	Path to sink s_{pq} that includes v_j and belongs to direction d_i
$\overline{P_{s_{pq}v_jd_i}}$	Path to sink s_{pq} including v_j in every candidate direction except for d_i
$C_{dv_jd_i}$	Downstream capacitance of the allowed interval of via vj for the paths $P_{s_{pq}v_jd_i}$
$C_{dv_j\overline{d_i}}$	Downstream capacitance of the allowed interval of via vj for the paths $\overline{P_{s_{pq}v_jd_i}}$

Different objective functions can be applied to optimize the performance of an interconnect structure. In this chapter, the weighted summation of the distributed Elmore delay of the branches of an interconnect tree is considered as the objective function,

$$T_w = \sum_{\forall s_{pq}} w_{s_{pq}} T_{s_{pq}}, \tag{8-1}$$

where w_{spq} and T_{spq} are the weight and distributed Elmore delay of sink s_{pq}, respectively. The weights are assigned to the sinks according to the criticality of the sinks. For a via connecting multiple interconnect segments, or equivalently, for a via with degree greater than two, there are several candidate directions d_i's along which the delay can be decreased. The placement of vias along these directions is constrained by the lengths l_{id}'s, as shown in Figure 8-1b, where the l_{di}'s are not generally equal. In addition, vias can span more than one physical plane. For example, consider the via connecting sinks s_{23} and s_{33}. This via traverses two physical planes, where the allowed interval for placing the via can be different for each plane.

Three different types of moves for an interplane via are defined. A *type*-1 move is shown in Figure 8-2a. This type of move requires the insertion of an intraplane via (to preserve connectivity), as depicted by a dot in Figure 8-2a. In the following analysis, the effect of these additional intraplane vias on the delay of the tree is assumed to be negligible. The impedance characteristics of the intraplane vias are assumed to be considerably lower than the impedance characteristics of the interplane vias [223], particularly if bulk CMOS technology is used for the upper planes. Alternatively, this effect can be included by appropriately shrinking the length of the allowed interval of the interplane via.

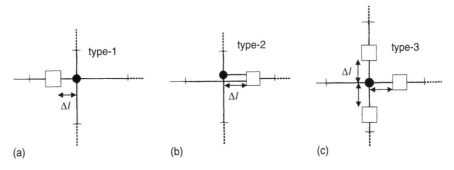

(a) (b) (c)

■ **FIGURE 8-2** Different interplane via moves: (a) *type-1* move (allowed), (b) *type-2* move (allowed), and (c) *type-3* move (prohibited).

A *type-2* move is shown in Figure 8-2b. A *type-2* move differs from a *type-1* move in that an additional interconnect segment of length Δl is inserted. Although an additional interconnect segment is required for this type of move, a reduction in the delay of the tree can occur. The segments of length Δl illustrated in Figure 8-2b are located on the same *y*-coordinate but on different physical planes, and yet are shown on different coordinates for added clarity.

Another type of move is illustrated in Figure 8-2c where additional interplane vias are inserted and is denoted as a *type-3* move. This type of move is not permitted for two reasons. The additional interplane vias outweigh the reduction in delay resulting from optimizing the length of the connected segments due to the high-impedance characteristics of the interplane vias. Additional interplane vias also increase the vertical interconnect density, which is undesirable. The routing congestion also increases as these vias typically block the metal layers within a plane, adversely affecting the length of the allowed intervals for the remaining nets.

As with two-terminal nets, certain constraints on the total length of each source to sink path apply to interconnect trees,

$$L_{s_{pq}} = l_1 + l_{v1} + \ldots l_j + l_{v_j} + \ldots + l_{v_{n-1}} + l_n, \qquad (8\text{-}2)$$

where l_j and l_{vj} are the length of the horizontal segment j and via v_j, respectively, which describe the path from the root of the tree to the sink s_{pq}. The number of segments that constitute this sink is denoted as n. The constraint in (8-2) is adapted to consider any increase in wirelength that can result from a *type-2* move for some branches of the tree. In addition, the length of each segment of the tree and the via placement are also constrained by

$$l_{j\min} \leq l_j \leq l_{di.j-1} + l_{j\min} + l_{di.j}, \forall i \in \{w, e, s, n\}, \forall j \in [2, n-1] \\ \text{and } l_{di.0} = l_{di.n} = 0, \qquad (8\text{-}3)$$

$$0 \leq x_j \leq l_{di.j}, \forall i \in \{w, e, s, n\}. \qquad (8\text{-}4)$$

Consequently, the constrained optimization problem for placing a via within an interplane interconnect tree can be described as

(P1) **minimize** T_w,

subject to (8-2)-(8-4), $\forall$ sink s_{pq} and via v_j.

With similar reasoning as for two-terminal nets, (P1) typically includes an indefinite quadratic form $\mathbf{l}^T \mathbf{A} \mathbf{l}$, where $\mathbf{A}$ is the matrix described in (7-20) adapted for interconnect trees. Certain transformations can be applied to convert (P1) into a convex optimization problem [231]; the objective functions, however, are no longer quadratic. Alternatively, accurate heuristics are described in the following section to determine the location of the vias that minimizes the delay of the interplane interconnect trees.

8.2 MULTITERMINAL INTERCONNECT VIA PLACEMENT HEURISTICS

Near-optimal heuristics for placing vias in interconnect trees in 3-D circuits are presented in this section. Initially, a heuristic for minimizing the weighted delay of the sinks of an interplane tree is described in Section 8.2.1. A second heuristic for optimizing the delay of a critical path in an interplane tree is discussed in Section 8.2.2.

8.2.1 Interconnect Trees

In this section, placing an interplane via within an interconnect tree in a 3-D circuit to minimize the summation of the weighted Elmore delay of the branches of the tree is investigated. Since several moves for the interplane vias are possible, as discussed in Section 8.1, the expressions that determine the via location are different in multiterminal nets. To determine which type of move for those vias with a connectivity degree greater than two will yield a decrease in the delay of the tree, the following conditions apply.

Condition 1: If $r_j > r_{j+1}$, only a *type-1* move for v_j can reduce the delay of the tree.

Proof: The proposition is analytically proven in Appendix D. The condition can also be intuitively explained. A *type-2* move increases by Δl the length of segment j. The reduction in l_{j+1} is counterbalanced by the additional segment with length Δl on the $j+1$ plane (see Figure 8-2b). Consequently, the total capacitance of the tree increases. If *condition 1* is satisfied, a *type-2* move also increases the total resistance of the tree and, therefore, the delay of the tree will only increase by this via move.

Condition 2: For a candidate direction d_i, if $r_j < r_{j+1}$ and

$$\sum_{\forall s_{pq} \in P_{s_{pq}v_jd_i}} w_{s_p}(r_j + r_{j+1})C_{dv_j\overline{d_i}} \leq \sum_{\forall s_{pq} \in P_{s_{pq}v_jd_i}} w_{s_p}(r_{j+1} - r_j)C_{dv_jd_i}, \qquad (8-5)$$

is satisfied, a *type-2* move can reduce the delay of the tree.

Proof: This condition is also intuitively demonstrated. All of the interconnect segments located upstream from v_j see an increase in the capacitance by $c_j \Delta l$, increasing the delay of each downstream sink v_j.

Consequently, only a reduction in the resistance can decrease the delay of the tree. Alternatively, the sinks located downstream of via v_j on the candidate direction di see a reduction in the upstream resistance by $(r_j - r_{j+1})\Delta l < 0$, while the sinks downstream of via v_j in the other directions see an increase in the upstream resistance by $(r_j + r_{j+1})\Delta l$. For a *type-2* move, resulting in a decrease in the delay of the tree, both the weighted sum of these two components as determined by the weight of the sinks and the downstream capacitances must be negative.

Condition 2 is evaluated for each via of a tree with degree greater than two. If (8-5) is satisfied for more than one direction, the direction that produces the greatest value of the RHS of (8-5) is considered the optimum direction for that via. Finally, note that both *conditions 1 and 2* are only necessary and not sufficient conditions. Following the notation listed in Table 8-1, the critical point for a via connecting two segments on planes j and $j + 1$ and satisfying *condition 1* is

$$x_{type-1} = \left[\frac{\left(\sum_{v_i \in U_{1_j}} \sum_{s_m \in \overline{P}_{s_m} U_{ij}} w_{s_m} R_{u_{ij}} + \sum_{s_p \in P_{s_p v_j}} w_{s_p} R_{u_j} \right)(c_{j+1} - c_j) - l_{v_j}(r_j c_{v_j} - r_{v_j} c_{j+1})}{\sum_{s_p \in P_{s_p v_j}} w_{s_p}(r_j c_j + r_{j+1} c_{j+1} - 2 r_j c_{j+1})} + \frac{(r_j - r_{j+1})(c_{j+1} l_{d_w} + C_{dv_j})}{\sum_{s_p \in P_{s_p v_j}} w_{s_p}(r_j c_j + r_{j+1} c_{j+1} - 2 r_j c_{j+1})} \right].$$

$$(8\text{-}6)$$

For a *type-2* move along a candidate direction d_i, the critical point for a via connecting two segments on planes j and $j + 1$ is

$$x_{type-2} = \left[\frac{\sum_{s_p \in P_{s_p v_j di}} w_{s_p} r_{j+1}(C_{dv_j d_i} + c_{j+1} l_{d_i}) - \sum_{v_i \in U_{1_j}} \sum_{s_m \in \overline{P}_{s_m} U_{ij}} w_{s_m} R_{u_{ij}} C_k}{\sum_{s_p \in P_{s_p v_j}} w_{s_p}(r_j c_j + r_{j+1} c_{j+1})} - \frac{\sum_{s_p \in P_{s_p v_j}} w_{s_p}(r_j c_{v_j} - c_{j+1} l_{d_i} + C_{dj} + r_{j+1} C_{dv_j \overline{di}} + R_{u_j} C_k)}{\sum_{s_p \in P_{s_p v_j}} w_{s_p}(r_j c_j + r_{j+1} c_{j+1})} \right].$$

$$(8\text{-}7)$$

8.2.2 Single Critical Sink Interconnect Trees

There are cases where the delay of only one branch of a tree is required to be optimized. Although the heuristic presented in the previous section can be used for this type of tree, a computationally simpler, yet accurate, optimization procedure for single critical net trees is described here. Denoting by s_c the critical sink of the tree, the weight for this sink w_{sc} is one, while the assigned weight for the remaining sinks is zero. Consequently, the expression that minimizes the delay is significantly simplified. In addition, the approach is different as compared to the optimization problem discussed in the previous section. More specifically, those interplane vias that belong to the critical branch (the on path vias) are placed according to the heuristic for two-terminal nets. There is no need to test *conditions 1* and *2* for these vias, as any *type-2* move only occurs in the direction that includes the critical sink. Regarding those vias that are not part of the critical path (the off path vias), these vias are placed to minimize the capacitance of the tree. A simple interconnect tree depicting this terminology is illustrated in Figure 8-3. This situation occurs because the noncritical sinks of the tree only contribute as capacitive loads to the delay of the critical sink.

The location of the off path vias is readily determined since the impedance characteristics of the interconnect segments are known. Note that in this sense, the placement of the off path vias is always optimal. Any loss of optimality is due to the location of the on path vias. As the near-optimal two-terminal net heuristic is used to place the on path vias, the loss of optimality is negligible. In the following section, these heuristics are used to develop efficient algorithms for placing vias in multiterminal nets in 3-D ICs.

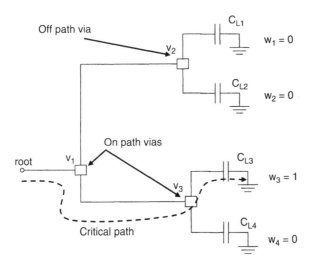

■ **FIGURE 8-3** Simple interconnect tree, illustrating a critical path ($w_3 = 1$) and on path and off path interplane vias.

8.3 **VIA PLACEMENT ALGORITHMS FOR INTERCONNECT TREES**

Efficient near-optimal algorithms for placing vias among interplane interconnects are investigated in this section. Based on the aforementioned heuristic described in Section 8.2.1, an efficient algorithm for interplane interconnect trees is presented in Section 8.3.1. A second algorithm that places interplane vias to minimize the delay for the particular case of interconnect trees with a single critical branch is discussed in Section 8.3.2.

8.3.1 **Interconnect Tree Via Placement Algorithm (ITVPA)**

The via placement optimization algorithm for multiterminal nets is presented in this section. The input to the algorithm is an interplane interconnect tree in which the minimum length of the segments, the weight of the sinks, and the length of the allowed intervals are provided. Pseudocode of the algorithm is shown in Figure 8-4. Due to the different types of moves that are possible in interplane interconnect trees, the candidate direction for via placement is initially determined in steps 1 through 5. The *move_type* routine operates from a leaf to the root, where the type and direction of the move of each via of the tree have degree greater than two. *Conditions 1* and *2* are tested for each via and direction. For those vias at the last level of the tree (close to the sinks), the downstream capacitance is determined. Alternatively, for those vias that belong to the next level closer to the root, the downstream capacitance cannot be accurately determined, and (8-5) is evaluated for the extreme values of C_{dj}. If (8-5) is satisfied for only one of these extrema, a via can be placed along a nonoptimal direction, resulting in loss of optimality. These instances, however, are not typically encountered, as discussed in the following section. In step 6, the *optimize_tree_delay* routine places the vias within a tree to minimize (8-1). This routine is a slight modification of the algorithm used for two-terminal nets.

■ **FIGURE 8-4** Pseudocode of the Interconnect Tree Via Placement Algorithm (ITVPA).

```
Interconnect Tree Via Placement Algorithm: (l_min, Δx, l_di, w_si)
1.  foreach physical plane i, i = n → 1
2.     foreach interplane via j on plane i
3.        if via degree > 2
4.           move_type(j)
        else
5.           goto step 2
6.  optimize_tree_delay()
7.  exit
```

8.3.2 **Single Critical Sink Interconnect Tree Via Placement Algorithm (SCSVPA)**

Although the heuristic presented in Section 8.2.2 can be used to improve the delay of trees with a single critical path, a simpler optimization procedure for single critical net trees is described in this section. The input to the single critical sink via placement algorithm (SCSVPA) is a description of the interplane interconnect tree where the minimum length of the segments, the weight of the sinks, and the length of the allowed intervals are provided. Pseudocode of the algorithm is shown in Figure 8-5. In steps 1 to 3, each of the off path vias is placed at the minimum capacitance location within the corresponding allowed interval. The direction, which includes the critical sink of the tree, is set by the *direction_move* routine as the direction along which the on path vias can be placed. In step 5, the *optimize_tree_delay* routine is utilized to determine the location of the on path vias. As previously mentioned, any loss of optimality for this type of tree results from the heuristic that places the via in two-terminal nets. As described in Chapter 7, this heuristic produces results similar to optimization solvers. SCSVPA naturally exhibits significantly lower computational time as compared to general purpose solvers.

8.4 **VIA PLACEMENT RESULTS AND DISCUSSION**

These algorithms are applied to several example interplane interconnect trees to evaluate efficiency and accuracy. Trees for different numbers of planes and sinks are analyzed. A discussion of the limitations of these algorithms is provided in this section. The impedance characteristics of the horizontal segments and vias are extracted for several interconnect structures using a commercial impedance extraction tool [224]. Copper interconnect is assumed with an effective resistivity of 2.2 $\mu\Omega$-cm. Based on the extracted impedances, the resistance and capacitance of the horizontal segments range from 25 Ω/mm to 125 Ω/mm and 100 fF/mm to 300 fF/mm, respectively, for a 90-nm CMOS technology node [233], [234]. The cross section of

Single Critical Sink Via Placement Algorithm: ($\mathbf{1_{min}}$, $\mathbf{\Delta x}$, $\mathbf{1_{di}}$, $\mathbf{w_{spq}}$)
1. foreach off path via *j*
2.   set via *j* to min. capacitance location
3. foreach on path via *i*
4.   direction_move(i)
5. optimize_tree_delay()
6. exit

■ **FIGURE 8-5** Pseudocode of the near-optimal Single Critical Sink interconnect tree Via Placement Algorithm (SCSVPA).

the vias is 1 μm × 1 μm, with 1 μm spacing from the surrounding horizontal metal layers, assuming an SOI process as described in [140]. The total and minimum length of each horizontal segment is randomly generated for each of the interconnect structures. For simplicity, all of the vias connect the segments of two adjacent physical planes. The savings in delay that can be achieved by optimally placing the vias are listed in Table 8-2 for different via placement scenarios.

The optimization results for interconnect trees with various numbers of sinks and planes are reported in Table 8-2. The accuracy and efficiency of ITVPA are similar to those of TTVPA (discussed in Chapter 7), as the optimization routine for ITVPA is the same as in TTVPA, after a move type and direction have correctly been determined. As mentioned in Section 8.3.1, a nonoptimal direction can be selected to place a via. A nonoptimal direction for via placement, however, can only be chosen if the connected branches are slightly asymmetric (*i.e.*, have similar impedance characteristics and criticality). In these cases, the value of the downstream capacitance and weight of these branches are close, making the weighted delay of these sinks similar. For such slightly asymmetric branches, (8-7) yields $x^*_{type\text{-}2} = 0$ for *type-2* moves, meaning that if a nonoptimal direction is chosen, the delay of the tree is not affected. Alternatively, a *type-2* move usually occurs for highly asymmetric trees where the delay of a branch dominates the delay of the tree. As described in (8-5), the type of move for an interplane via depends on the weight of the branches and the impedance characteristics of the interconnect segments. Consider the symmetric tree shown in Figure 8-6. The difference in the criticality and impedance of the branches is captured from the value of the assigned weights.

In Table 8-3, the optimum location for via v_2 is listed for various weights. From these values, note that a *type-2* move for via v_2 occurs only when branch s_2 dominates the delay of the tree. When the assigned weights of the branches of the tree are of similar value, a *type-2* move does not occur, as this move would only increase the

■ **FIGURE 8-6** A symmetric tree including two interplane vias. The interconnect parameters are $r_1 = 10.98\ \Omega/\text{mm}$, $r_2 = 11.97\ \Omega/\text{mm}$, $r_3 = 96.31\ \Omega/\text{mm}$, $c_1 = 147.89\ \text{fF/mm}$, $c_2 = 202$ fF/mm, $c_3 = 388.51\ \text{fF/mm}$, and allowed interval $l_{d,v2} = 75\ \mu\text{m}$.

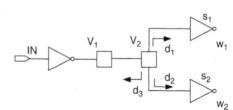

Table 8-2 Optimization results for various interplane interconnect trees for different number of sinks and physical planes *n*.

n	Number of Sinks	Avg. Branch Length [μm]	Avg. Maximum Branch Length [μm]	l_{di}'s [μm]	Delay Improvement [%]				Instances
					$x_i^* = l_{di}/2$		$x_i^* = 0$		
					Avg	Max	Avg	Max	
3	4	153	186	50	2.72	9.33	3.79	11.25	10000
3	4	307	376	100	4.23	15.17	6.03	17.94	10000
4	4	208	273	50	1.11	3.53	2.49	5.63	5000
4	4	828	1100	200	3.12	10.29	6.42	13.50	5000
4	4	1243	1650	300	4.07	14.15	7.76	19.38	5000
4	8	431	569	100	3.90	13.24	7.71	19.68	10238
5	4	264	362	50	1.25	3.83	2.40	5.89	5000
5	4	1054	1452	200	3.62	11.55	6.56	12.04	5000
5	4	791	1089	300	3.90	11.61	6.95	19.34	5000
5	8	454	660	50	0.90	2.69	2.27	4.98	5000
5	8	521	738	100	1.78	5.55	4.33	8.40	5000
5	8	779	1111	150	2.38	7.44	5.67	11.90	5000
5	8	1038	1481	200	2.91	8.71	6.74	12.58	5000
6	8	306	455	50	1.11	3.17	2.36	4.89	5000
6	8	615	913	100	2.00	5.44	4.09	9.85	5000
6	8	922	1373	150	2.72	7.01	5.43	11.72	5000
6	8	921	1371	200	3.32	10.02	6.61	14.21	5000
6	16	555	845	50	0.86	2.74	2.52	4.95	4970
6	16	637	934	100	1.68	4.82	4.84	9.26	5059
6	16	953	1404	150	2.28	6.10	6.32	12.96	5021

interconnect delay of the tree as determined by the ITVPA (*i.e.*, $x_j^* = \Delta x_j$). Consequently, if the weight of the sinks, which originate from a via, are similar for a *type-2* move, the algorithm does not allow the via to be relocated.

Table 8-3 Optimal via location, direction of move, and type of move for via v_2 shown in Figure 8-6, as determined from ITVPA for various values of w_1 and w_2.

w_1	w_2	$x^*_{v_2}$ [μm]	Move	Direction, d_i	w_1	w_2	$x^*_{v_2}$ [μm]	Move	Direction, d_i
0.50	0.05	Δx_{v2}	type-1	d_0	0.45	0.55	Δx_{v2}	type-2	d_1
0.55	0.45	Δx_{v2}	type-1	d_0	0.40	0.60	Δx_{v2}	type-1	d_0
0.60	0.40	Δx_{v2}	type-2	d_1	0.35	0.65	Δx_{v2}	type-2	d_2
0.65	0.35	Δx_{v2}	type-2	d_1	0.30	0.70	Δx_{v2}	type-2	d_2
0.70	0.30	Δx_{v2}	type-2	d_1	0.25	0.75	Δx_{v2}	type-2	d_2
0.75	0.25	Δx_{v2}	type-2	d_1	0.20	0.80	Δx_{v2}	type-2	d_2
0.80	0.20	Δx_{v2}	type-2	d_1	0.15	0.85	Δx_{v2}	type-2	d_2
0.85	0.15	Δx_{v2}	type-2	d_1	0.10	0.90	Δx_{v2}	type-2	d_2
0.90	0.10	Δx_{v2}	type-2	d_1	0.05	0.95	Δx_{v2}	type-2	d_2
0.95	0.05	Δx_{v2}	type-2	d_1	0.03	0.97	68.70	type-2	d_2

The improvement in delay of the interconnect trees is listed in columns 6 through 9 of Table 8-2. The results are compared to the case where the vias are initially placed at the center of the allowed interval (*i.e.*, $x_i = l_{di} / 2$) and to the case where the vias are placed at the lower edge of the allowed interval (*i.e.*, $x_i = 0$). The improvement in delay depends on the length of the allowed interval. This dependence, however, is weak as compared to two-terminal nets. In addition, the improvement in delay is lower than point-to-point nets for the same allowed length intervals. This reduction in delay improvement occurs for two reasons.

For those vias with degree greater than two, which constitute the majority of interplane vias in interconnect trees, after the type of move for each via is determined, the actual interval length that these vias are allowed to move is $l_{di} / 2$ and not l_{di} (see Figure 8-2). Furthermore, in ITVPA, any modifications to the routing tree are strictly confined within the allowed interval to least affect the routing tree.

This constraint requires an additional interconnect segment for *type-2* moves. If this constraint is relaxed, an additional interconnect segment is not necessary, and the length of the interconnect segments can be further reduced, resulting in a considerably greater improvement in speed. Maintaining fixed paths limits the efficiency of the algorithms; however, these algorithms are applicable to placement and routing tools for 3-D ICs as long as these tools provide an allowed interval for via placement. In addition, interconnect routing can consider other important design objectives such as thermal effects or routing congestion. These algorithms for placing vias in multiterminal nets can be applied as a subsequent postprocessing step without significantly affecting the initial layout produced by existing tools.

In Table 8-4, placement results for single critical branch interconnect trees are reported. The improvement in the delay of these trees is listed in columns 6 through 9 of Table 8-4, as compared to the situation where the vias are initially placed at the center of the allowed interval (*i.e.*, $x_i = l_{di} / 2$) and where the vias are placed at the lower edge of the allowed interval (*i.e.*, $x_i = 0$). This improvement is lower than for those interconnect trees listed in Table 8-4, as only *type-1* moves can occur for the off path vias. Indeed, any *type-2* move for the off path via only increases the off path capacitance and, in turn, the delay of the critical leaf. A smaller number of vias can, therefore, be relocated to reduce the delay of the single critical sink trees. Alternatively, for off path vias, placing the via at the center of the allowed intervals usually produces an optimum placement, resulting in a smaller overall improvement in the delay of this type of tree, as is demonstrated by the test structures. This situation occurs because *type-2* moves for the off path vias are not allowed since a *type-2* move results in an increase in the off path capacitance.

Typically, the larger the allowed interval, the greater the improvement in delay. Consequently, efficient placement tools for 3-D circuits that generate sufficiently large allowed intervals are desirable. These intervals can be available space reserved for interplane interconnect routing. For interconnect trees, the improvement in delay is smaller than for two-terminal nets. This decreased improvement in delay is due to the constraint of placing the vias within the allowed intervals to minimally affect the local routing congestion. If the placement of the vias is permitted within an entire region (*e.g.*, the bounding box of the net), a greater decrease in delay can occur. Assigning such a region for placing vias, however, increases the congestion within a 3-D circuit as the same number of vias compete for sparser routing resources.

Table 8-4 Optimization results for various single critical sink interconnect trees for different number of sinks and physical planes n.

n	Number of Sinks	Avg. Branch Length [μm]	Avg. Maximum Branch Length [μm]	Δx_i's [μm]	Delay Improvement [%]				Instances
					$x_i^* = l_{di}/2$		$x_i^* = 0$		
					Avg	Max	Avg	Max	
4	4	341	453	50	2.72	8.95	3.70	14.63	5000
4	4	1021	1363	150	1.61	5.52	2.18	11.01	5000
4	4	1368	1821	200	1.36	5.49	1.92	8.59	5000
5	4	433	595	50	3.09	9.37	4.55	19.44	5000
5	4	1299	1790	150	1.80	5.55	2.85	13.42	5000
5	4	1734	2391	200	1.51	5.66	2.44	11.35	5000
5	8	427	612	50	2.53	8.10	4.09	16.20	5000
5	8	853	1227	100	1.94	7.22	2.98	12.63	5000
5	8	1282	1845	150	1.57	6.55	2.46	14.04	5000
5	8	1711	2461	200	1.33	4.81	2.25	10.44	5000
6	8	505	753	50	2.88	8.90	4.39	17.8	5000
6	8	1009	1512	100	2.14	6.10	3.45	15.20	5000
6	8	1511	2265	150	1.71	5.46	2.83	12.16	5000
6	16	523	779	50	2.52	8.29	4.54	13.25	4963
6	16	1045	1564	100	1.91	6.69	3.10	10.53	4977
6	16	1563	2351	150	1.55	5.36	2.96	12.37	4976

Despite the considerably lower computational time of these algorithms, further speed improvements can be achieved if more than one net are simultaneously processed. Although these algorithms support multiple net optimization without significant modification, a single net at a time approach likely yields improved results as the most critical nets are routed first. Net ordering algorithms [20] can be used to prioritize the routing of these interconnects, permitting the delay of these nets to be considerably reduced. In addition, since the number of interplane interconnects is small as compared to the number of intraplane interconnects [123], processing these interconnects one net at a time will not significantly increase the total computational time.

Thermal issues are important in 3-D ICs, as discussed in Chapter 6, where additional dummy vias are utilized to control the average and peak temperature of the upper planes within a 3-D system. In addition, thermally aware cell placement improves the heat distribution and removal characteristics. These two techniques are decoupled from the via placement problem, which is considered a later step in the design process. Consequently, thermal issues are not strongly connected with the via placement approach.

In the techniques discussed in Chapter 6, the thermal vias are placed in the available space among the blocks with either uniform or non-uniform densities. Alternatively, some of the thermal vias within this available space can be replaced with signal vias connecting circuits on different planes. Placing the signal vias prior to placing the thermal vias within these regions results in large allowed intervals, thereby improving the effectiveness of this via placement technique. In this case, where the via placement can significantly enhance the thermal profile of a physical plane, the allowed interval for some vias can be decreased or removed. This practice, however, trades off performance for thermal management.

8.5 SUMMARY

Algorithms for the timing-driven placement of interplane interconnect trees are presented in this chapter. The primary characteristics of these algorithms are:

- Several types of moves exist for interplane vias, requiring the insertion of either intraplane or interplane vias. Moves that require additional interplane vias are excluded to maintain a low interplane via density.

- The weighted summation of the delay at the sinks of a tree is the objective function that is minimized.

- A two-step heuristic is presented. The direction of the move for each via is initially determined followed by the second step, where the via is placed to minimize the delay.

- Another heuristic for placing interplane vias within trees to minimize the delay of the critical path is provided.

- Both of these heuristics exhibit linear complexity with the number of interplane vias.

- The improvement in delay depends upon the size of the allowed intervals for placing each via.

3-D Circuit Architectures

Technological, physical, and thermal design methodologies have been presented in the previous chapters. The architectural implications of adding the third dimension in the integrated circuit design process are discussed in this chapter. Various wire-limited integrated systems are explored where the third dimension can mitigate many interconnect issues. Primary examples of this circuit category are the microprocessor-memory system, on-chip networks, and field programmable gate arrays (FPGAs). Although networks-on-chip (NoC) and FPGAs are generic communication fabrics as compared to microprocessors, the effects of the third dimension on the performance of a microprocessor are discussed here due to the importance of this circuit type.

The performance enhancements that originate from the 3-D implementation of wire-dominated circuits are presented in the following sections. These results are based on analytic models and academic design tools under development with exploratory capabilities. Existing performance limitations of these circuits are summarized in the following section, where a categorization of these circuits is also provided. Three-dimensional architectural choices and corresponding tradeoffs for microprocessors, memories, and microprocessor-memory systems are discussed in Section 9.2. Three-dimensional topologies for on-chip networks are presented and evaluated in Section 9.3. Both analytic expressions and simulation tools are utilized to explore these topologies. Finally, the extension to the third dimension of an important design solution, namely FPGAs, is analyzed in Section 9.4. Related aids for the physical design of 3-D FPGAs are also discussed. A brief summary of the analysis of these 3-D architectures is offered in Section 9.5.

9.1 CLASSIFICATION OF WIRE-LIMITED 3-D CIRCUITS

Any two-dimensional integrated circuit can be vertically fabricated with one or more processes developed for 3-D circuits. The benefits, however, that stem from this implementation vary for different circuits [123]. Wire-dominated circuits are good candidates for vertical integration since these circuits greatly benefit from the significant decrease in wirelength. Performance projections as previously discussed in Chapter 4 highlight this situation. Consequently, only communication centric circuits are emphasized in this discussion. The different circuit categories considered in this chapter are illustrated in Figure 9-1.

The first category includes application-specific ICs (ASIC) that are typically part of a larger computing system. A criterion for a circuit to belong to this category is whether the third dimension can considerably improve the primary performance characteristics of a circuit, such as speed, power, and area. A fast Fourier transform circuit is an example of a 3-D ASIC [238]. Memory arrays and microprocessors are other circuit examples that belong to this category. 3-D integration is amenable to the interconnect structures connecting these components comprising more complex integrated systems.

Communication fabrics, such as networks-on-chip and FPGAs, are particularly appropriate for vertical integration. For instance, several low-latency and high-throughput multidimensional topologies have been presented in the past for traditional interconnection networks, such as three-dimensional meshes and tori [239]. These topologies, depicted in Figure 9-2, are not usually considered for on-chip networks due to

■ **FIGURE 9-1** Taxonomy of 3-D architectures for wire-limited circuits.

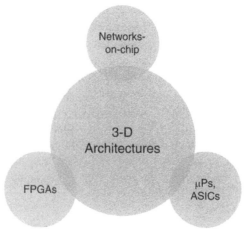

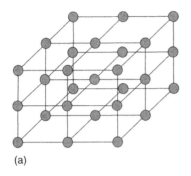 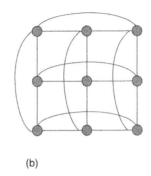

■ **FIGURE 9-2** Popular interconnection network topologies: (a) three-dimensional mesh and (b) two-dimensional torus.

(a)   (b)

the long interconnects that hamper the overall performance of the network. These limitations are naturally circumvented when a third degree of design freedom is added.

In the case of FPGAs, a design style for an increasing number of low-volume applications, vertical integration offers a twofold opportunity: increased interconnectivity among the logic blocks and shorter distances among these blocks. These advantages will enhance the performance of FPGAs, which is the major impediment of this design style. Consequently, 3-D architectures are indispensable for contemporary FPGAs. Each of these circuit categories is successively reviewed in the remainder of the chapter. Related design tools are also discussed, and further issues in the design of these circuits are highlighted.

9.2 THREE-DIMENSIONAL MICROPROCESSORS AND MEMORIES

Microprocessor and memory circuits constitute a fundamental component of every computing system. Due to the use of these circuits in myriad applications, the effect of 3-D integration on this type of systems is of significant interest. Both of these types of circuits are amenable to a variety of 3-D architectural alternatives. The partitioning scheme used on standard 2-D circuits drastically affects the characteristics of the resulting 3-D architectures. The different partitioning levels and related building elements based on these partitions are illustrated in Figures 9-3 and 9-4, respectively. A finer partitioning level typically requires a larger design effort and higher vertical interconnect densities. Consequently, each partitioning level is only compatible with a specific 3-D technology. Note that the intention here is not to determine an effective partitioning methodology specific to 3-D microprocessors

■ **FIGURE 9-3** Different partitioning levels and related design complexity vs. the architectural granularity for 3-D microprocessors.

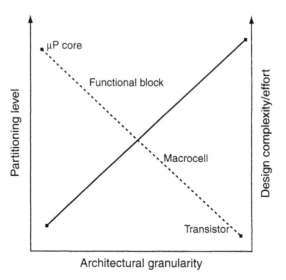

■ **FIGURE 9-4** An example of different partition levels for a 3-D microprocessor system at the (a) core, (b) functional unit block (FUB), (c) macrocell, and (d) transistor level.

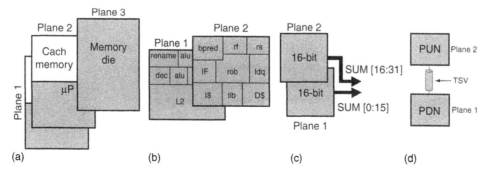

and memories but rather to determine the architectural granularity that most improves the performance of these circuits. The physical design techniques described in Chapter 5 can be used to achieve the optimum design objectives for a particular architecture.

Different architectures for several blocks not including the on-chip cache memory of a microprocessor are discussed in Section 9.2.1. The 3-D organization of the cache memory is described in Section 9.2.2. Finally, the 3-D integration of the combined microprocessor and memory is discussed in Section 9.2.3.

9.2.1 Three-Dimensional Microprocessor Logic Blocks

The microprocessor circuit discussed here consists of one logic core and an on-chip cache memory. The architectures can be extended to multicore microprocessor systems. For a microprocessor circuit, partitioning the functional blocks or macrocells (*i.e.*, circuits within the functional blocks) to several planes (see Figure 9-4) is meaningful and can improve the performance of the individual functional blocks and the microprocessor. For a microprocessor system, partitioning at the core level is also possible. The resulting effects on a microprocessor are higher speed, additional instructions per cycle (IPC), and a decrease in the number of pipeline stages.

In general, the 3-D implementation of a wire-limited functional block (*i.e.*, partitioning at the macrocell level) decreases the dissipated power and delay of the blocks. The benefits, however, are minimal for blocks that do not include relatively long wires. In this case, some performance improvement exists due to the decrease in the area of the block and, consequently, in the length of the wires traversing the block. In addition to the cache memory, many other blocks within the processor core can be usefully implemented on more than one physical plane. These blocks include, for example, the instruction scheduler, arithmetic circuits, content addressable memories, and register files.

The instruction scheduler is a critical component constraining the maximum clock frequency and dissipates considerable power [240]. Placing this block in two planes results in lower delay and power by, respectively, 44% and 16% [241]. Additional savings in both delay and power can be achieved by using three or four planes; however, the savings saturate rapidly. This situation is more pronounced in the case of arithmetic units, such as adders and logarithmic shifters. Delay and power improvements for Brent Kung [242] and Kogge Stone [243] adders are listed in Table 9-1. As indicated from these results, the benefits of utilizing more than two planes are negligible. Further reductions in delay are not possible for more than two planes since the delay of the logic dominates the delay of the interconnect.

Consequently, partitioning at the macrocell level is not necessarily helpful for every microprocessor component and should be carefully applied to ensure that only the wire-dominated blocks are designed as 3-D blocks. Alternatively, another architectural approach does not split but simply stacks the functional blocks on adjacent physical planes, decreasing the length of the wires shared by these blocks. As an example of this approach, consider the two-plane 3-D design of the Intel Pentium® 4

Table 9-1 Performance and power improvements of 3-D over 2-D architectures [241].

| # of input bits | Kogge–Stone Adder | | | Brent–Kung Adder |
| | 16-bits | | 32-bits | 32-bits |
	Delay	Power	Delay	Delay
2 planes	20.23%	8%	9.6%	13.3%
3 planes	23.60%	15%	20.0%	18.1%
4 planes	32.70%	22%	20.0%	21.7%

processor where 25% of the pipeline stages in the 2-D architecture are eliminated, improving performance by almost 15% [244]. The power consumption is also decreased by almost 15%. A similar architectural approach for the Alpha 21364 [245] processor resulted in a 7.3% and 10.3% increase in the IPC for two and four planes, respectively [246].

A potential hurdle in the performance improvement is the introduction of new hot spots or an increase in the peak temperature as compared to a 2-D design of the microprocessor. Thermal analysis of the 3-D Intel Pentium® 4 has shown that the maximum temperature is only 2°C greater as compared to the 2-D counterpart, reaching a temperature of 101°C [246]. If maintaining the same thermal profile is the predominant objective, the power supply can be scaled, partially limiting the improvement in speed. For this specific example, voltage scaling pacified any temperature increase while providing an 8% performance improvement and 34% decrease in power (for two planes) [244]. In these case studies, a 2-D architecture has been redesigned into a 3-D microprocessor architecture. Greater enhancements can be realized if the microprocessor is designed from scratch, initially targeting a 3-D technology as compared to simply migrating a 2-D circuit into a multiplane stack. A considerable portion of the overall performance improvement in a microprocessor circuit can be achieved by distributing the cache memory onto several physical planes, as described in the following section.

9.2.2 **Three-Dimensional Design of Cache Memories**

Data exchange between the processor logic core and the memory has traditionally been a fundamental performance bottleneck. A small amount of memory, specifically cache memory, is therefore integrated with the logic circuitry offering very fast data transfer, while the majority of the main memory is implemented off-chip. The size and organization of the cache memory greatly depend on the architecture of the microprocessor and has steadily increased over the past several microprocessor generations [247], [248]. Due to the small size of the cache memory, a common problem is that data is often fetched from the main memory. This situation, widely known as a cache miss, is a high latency task. Increasing the cache memory size can partly lower the cache miss rate. Three-dimensional integration supports both larger and faster cache memories. The former characteristic is achieved by adding more memory on the upper planes of a 3-D stack, and the latter objective can be enhanced by constructing novel cache architectures with shorter interconnects.

A schematic view of a 2-D 32 KB cache is illustrated in Figure 9-5, where only the data array is depicted. The memory is arranged into smaller arrays (subarrays) to decrease both the access time and power dissipation. Each memory subarray i is denoted as Block i. The size of the subarrays is determined by two parameters, N_{dwl} and N_{dbl}, which correspond to the divisions of the initial number of word and bit lines, respectively. In this example, each subarray contains 128×256 bits. In addition to the SRAM arrays, other circuits are shown in Figure 9-5. The local word line decoders and drivers are placed on the left side of each subarray. The multiplexers and sense

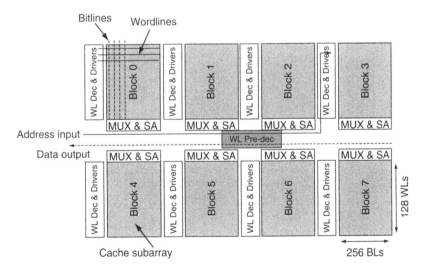

■**FIGURE 9-5** 2-D organization of a cache memory [249].

amplifiers are located on the bottom and top side of each row. The word line pre-decoder is placed at the center of the entire memory.

There are several ways to partition this structure into multiple planes. An example of a 2-D and 3-D organization of a 32 Kb cache memory is schematically illustrated in Figure 9-6. The memory can be stacked at the functional block, macrocell, and transistor levels, where a functional block is considered in this case to be equivalent to a memory subarray. Halving the memory and placing each half on a separate plane decreases the length of the global wires, such as the address input to the word line pre-decoder nets, the data output lines, and the wires used for synchronization. Partitioning at the functional block level does not improve the delay and power of the subarrays, which adds to the total access time and power dissipation.

Dividing each subarray can therefore result in lower access time and power consumption. Partitioning occurs along the x- and y-directions, halving the length of the bit and word lines, respectively, at each division. An example of a subarray partition is depicted in Figure 9-7. The number of partitions along each direction is characterized by the parameters N_x and N_y. For word line partitions ($N_x > 1$), the word lines are replicated on the upper planes, as shown in Figure 9-7. The length of the word lines, however, decreases, resulting in smaller device sizes within the drivers and local decoders. Furthermore, the area of the overall array decreases, leading to shorter global wires, such as the input address from the pre-decoder to the local decoders and the data output lines.

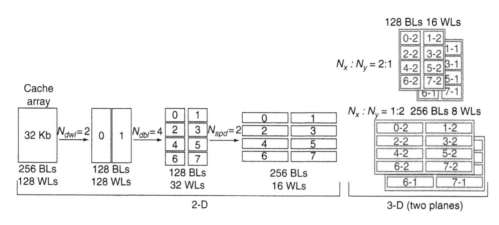

■ **FIGURE 9-6** Schematic of 2-D and 3-D organization of a 32-Kb cache memory array. N_{spd} is the number of sets connected to a word line.

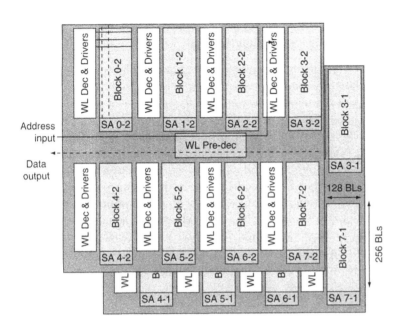

■ **FIGURE 9-7** Word line partitioning onto two planes of the 2-D cache memory shown in Figure 9-5 [246].

In the case of bit line partitioning ($N_y > 1$), the length of these lines and the number of pass transistors tied to each bit line are reduced, as illustrated in Figure 9-8. The sense amplifiers can either be replicated on the upper planes or shared among the bit lines from more than one plane. In the former case, the leakage current increases but the access time is improved, while in the latter case the power savings is

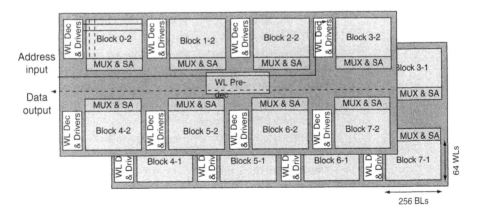

■ **FIGURE 9-8** Bit line partitioning onto two planes of the 2-D cache memory shown in Figure 9-5 [246].

greater. The speed enhancement, however, is not significant due to the requirement for bit multiplexing.

In general, word line partitioning results in a smaller delay but not necessarily a larger power savings. More specifically, for high-performance caches, partitioning the word lines offers greater savings in both delay and energy. For instance, a 1-MB cache where $N_x = 4$ and $N_y = 1$ is faster by 16.3% as compared to a partition where $N_x = 1$ and $N_y = 4$ [249]. Alternatively, bit line partitioning is more efficient for low-power memories. When the memory is designed for low power, bit line partitioning decreases the power by approximately 14% as compared to the reduction achieved by word line partitioning [249]. This behavior can be explained by considering the original 2-D design of the cache memory. High-performance memories favor wide arrays, which implies longer word lines, while low-power memories exhibit a greater height, resulting in longer bit lines.

Finally, partitioning is also possible at the transistor level where the basic six transistor SRAM cell can be split among the planes of the 3-D stack. This extra fine granularity, however, has a negative effect on the total area of the memory, since the size of a TSV is typically larger than the area of an SRAM cell [250]. Consequently, partitioning at the macrocell level offers the greatest advantages for 3-D cache memory architectures.

Analyzing the performance of these architectures is a multivariable task, and design aids to support this analysis are needed. PRACTICS [251] and 3-D CACTI [249] offer exploratory capabilities for cache memories. The cache cycle time models in both of these tools are based on CACTI, an exploratory tool for 2-D memories [252]. These tools utilize delay [253], energy [254], and thermal models [255] as well as extracted delay and power profiles from SPICE simulations to characterize a cache memory. Several parameters are treated as variables. The optimum value of these parameters is determined based on the primary design objective of the architecture, such as high speed or low power. Although these tools are not highly accurate, early architectural decisions can be explored. Choosing an appropriate 3-D architecture for the cache memory further increases the performance of the microprocessor in addition to performance improvements offered by the 3-D design of specific blocks within the processor. In this case, the primary limitation of the system is the off-chip main memory discussed in the following section.

9.2.3 **Architecting a 3-D Microprocessor — Memory System**

Although partitioning the cache memory on multiple planes enhances the performance of the microprocessor, data transfer to and from the main memory remains a significant hindrance. The ultimate 3-D solution is to stack the main memory on the upper planes of a 3-D microprocessor system. This option may be feasible for low-performance processors with low-memory requirements [256]. For modern computing systems with considerable memory demands, increasing the size of the on-chip cache, mainly the second level (L2) cache, is an efficient approach to improve performance [257]. Two systems based on two different microprocessors are considered; one approach contains a RISC processor [258], [259] while the other approach includes an Intel Core 2 Duo® processor [260].

To evaluate the effectiveness of the RISC system, the average time per instruction is a useful metric. For this system, the main memory is implemented within the 3-D stack. This practice offers a higher buss bandwidth between the main memory and the L2 cache, which, in turn, decreases the time required to access the main memory. The reduction in the average number of instructions, however, is small — about 6.1% [256]. In addition, stacking many memory planes within a 3-D stack can be technologically and thermally challenging.

A more practical approach is to increase the size of the L2 cache by utilizing a small number of either SRAM or DRAM planes on top of the processor. An Intel Core 2 Duo® system is illustrated in Figure 9-9, where various configurations of the cache memory are illustrated [244].

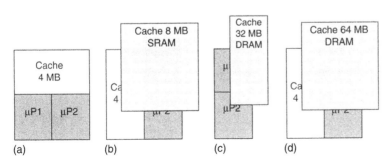

(a) (b) (c) (d)

■ **FIGURE** 9-9 Different organizations of a microprocessor system: (a) the 2-D baseline system, (b) a second plane with 8 MB SRAM cache memory, (c) a second plane with 32-MB SRAM cache memory, and (d) a second plane with 64-MB DRAM cache memory [244].

Note that in some of these configurations, both level one (L1) and L2 caches are included on the same plane. Increasing the size of the L2 cache increases the time to access this memory. The advantage of the reduced cache miss rates due to more data and instructions available on-chip considerably outweighs, however, the increase in access time. Each of the architectures illustrated in Figure 9-9 decreases the number of cycles per memory access for a large number of benchmarks [244]. The only exceptions are those benchmarks where the 4-MB memory in the baseline implementation is sufficient. In addition, the power consumed by the microprocessor-memory system decreases since a smaller number of transactions takes place over the off-chip high-capacitive buss. Furthermore, the required bandwidth of the off-chip buss drops as much as three times as compared to a 2-D system [244].

The presence of a second memory plane naturally increases the on-chip power consumption as compared to a 2-D microprocessor. The estimated power consumed by each configuration, shown in Figure 9-9, is listed in Table 9-2. Despite the increase in power dissipation, the highest increase in temperature as compared to the baseline system does not exceed 5°C, with a maximum (and manageable) temperature of 92.9°C [244] [261].

Larger on-chip memories can decrease cache miss rates for single- or double-core microprocessors [262]. For large-scale systems with tens of cores, however, memory access time and the related buss bandwidth can be a performance bottleneck. An on-chip network is an effective way to overcome these issues. Therefore, 3-D architectures for networks-on-chip are the subject of the following section.

Table 9-2 Power dissipation of the 3-D microprocessor architectures [244].

Architecture	Power Consumption [W]		
	Plane 1	Plane 2	Total
Figure 9-9a (2-D)	92	-	92
Figure 9-9b	92	14	106
Figure 9-9c	85	3.1	88.1
Figure 9-9d	92	6.2	98.2

9.3 THREE-DIMENSIONAL NETWORKS-ON-CHIP

Networks-on-chip is a developing design paradigm to enhance interconnections within complex integrated systems. These networks have an interconnect structure that provides Internet-like communication among various elements of the network; however, on-chip networks differ from traditional interconnection networks in that communication among the network elements is implemented through the on-chip routing layers rather than the metal tracks of the package or printed circuit board (PCB).

Networks-on-chip offer high flexibility and regularity, supporting simpler interconnect models and greater fault tolerance. The canonical interconnect backbone of the network combined with appropriate communication protocols enhances the flexibility of these systems [30]. An NoC provides communication among a variety of functional intellectual property (IP) blocks or processing elements (PE), such as processor and DSP cores, memory blocks, FPGA blocks, and dedicated hardware, serving a plethora of applications that include image processing, personal devices, and mobile handsets, [263]–[265] (the terms IP block and PEs are interchangeably used in this chapter to describe functional structures connected by an NoC). The intra-PE delay, however, cannot be reduced by the network. Furthermore, the length of the communication channel is primarily determined by the area of the PE, which is typically unaffected by the network structure. By merging vertical integration with NoC, many of the individual limitations of 3-D ICs and NoC are circumvented, yielding a robust design paradigm with unprecedented capabilities.

Research in 3-D NoC is only now emerging [206], [266]–[268]. Addo-Quaye [206] recently presented an algorithm for the thermal-aware mapping and placement of 3-D NoC, including regular mesh topologies. Li *et al.* [268] proposed a similar 3-D NoC topology employing a buss structure for communicating among PEs located on different physical planes. Targeting multiprocessor systems, the proposed scheme in [268] considerably reduces cache latencies by utilizing the third dimension. Multidimensional interconnection networks have been studied under various constraints, such as constant bisection-width and pin-out constraints [239]. Networks on-chip differ from generic interconnection networks, however, in that NoCs are not limited by the channel width or pin-out. Alternatively, physical constraints specific to 3-D NoC, such as the number of nodes that can be implemented in the third dimension and the asymmetry in the length of the channels of the network, have to be considered.

In this chapter, various possible topologies for 3-D NoC are presented. In addition, analytic models for the zero-load latency and power consumption with delay constraints of these networks that capture the effects of the topology on the performance of 3-D NoC are described. Optimum topologies are shown to exist that minimize the zero-load latency and power consumption of a network. These optimum topologies depend on a number of parameters characterizing both the router and the communication channel, such as the number of ports of the network, the length of the communication channel, and the impedance characteristics of the interconnect. Various tradeoffs among these parameters that determine the minimum latency and power consumption topology of a network are investigated for different network sizes. A cycle-accurate simulator for 3-D topologies is also discussed. This tool is used to investigate the behavior of several 3-D topologies under broad traffic scenarios.

Several interesting topologies, which are the topic of this chapter, emerge by incorporating the third dimension in NoC. In the following section, various topological choices for 3-D NoC are reviewed. In Section 9.3.2, an analytic model of the zero-load latency of traditional interconnection networks is adapted for each of the proposed 3-D NoC topologies, while the power consumption model of these network topologies is described in Section 9.3.3. In Section 9.3.4, the 3-D NoC topologies are compared in terms of the zero-load network latency and power consumption with delay constraints, and guidelines for the optimum design of speed-driven or power-driven NoC structures are provided. An advanced NoC simulator, which is used to evaluate the performance of a broad variety of 3-D network topologies, is presented in Section 9.3.5.

9.3.1 **3-D NoC Topologies**

Various topologies for 3-D networks are presented, and related terminology is introduced in this section. Mesh structures have been a popular network topology for conventional 2-D NoC [269], [270]. A fundamental element of a mesh network is illustrated in Figure 9-10a, where each processing element (PE) is connected to the network through a router. A PE can be integrated either on a single physical plane (2-D IC) or on several physical planes (3-D IC). Each router in a 2-D NoC is connected to a neighboring router in one of four directions. Consequently, each router has five ports. Alternatively, in a 3-D NoC, the router typically connects to two additional neighboring routers located on the adjacent physical planes. The architecture of the router

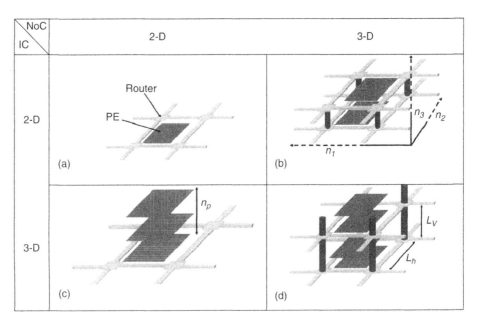

■ **FIGURE 9-10** Various NoC topologies (not to scale): (a) 2-D IC – 2-D NoC, (b) 2-D IC – 3-D NoC, (c) 3-D IC – 2-D NoC, and (d) 3-D IC – 3-D NoC.

is considered here to be a canonical router with input and output buffering [271]. The combination of a PE and router is called a network node. For a 2-D mesh network, the total number of nodes N is $N = n_1 \times n_2$, where n_i is the number of nodes included in the i_{th} physical dimension.

Integration in the third dimension introduces a variety of topological choices for NoCs. For a 3-D NoC as shown in Figure 9-10b, the total number of nodes is $N = n_1 \times n_2 \times n_3$, where n_3 is the number of nodes in the third dimension. In this topology, each PE is on a single yet possibly different physical plane (2-D IC – 3-D NoC). Alternatively, a PE can be implemented on only one of the n_3 physical planes of the system. The 3-D system therefore contains $n_1 \times n_2$ PEs on each one of the n_3 physical planes where the total number of nodes is N. This topology is discussed in [206] and [268]. A 3-D NoC topology is illustrated in Figure 9-10c, where the interconnect network is contained within one physical plane (*i.e.*, $n_3 = 1$), while each PE is integrated on multiple planes, notated as n_p (3-D IC – 2-D NoC) [267]. Finally, a hybrid 3-D NoC based on the two previous topologies is depicted in Figure 9-10d [267]. In this NoC topology, both the interconnect network and the PEs can span more than one physical plane of the stack (3-D IC – 3-D NoC). In the following section, latency expressions for each of the NoC topologies are described, assuming a zero-load model.

9.3.2 Zero-Load Latency for 3-D NoC

In this section, analytic models of the zero-load latency of each of the 3-D NoC topologies are described. The zero-load network latency is widely used as a performance metric in traditional interconnection networks [272]. The zero-load latency of a network is the latency where only one packet traverses the network. Although such a model does not consider contention among packets, the zero-load latency model can be used to describe the effect of a topology on the performance of a network. The zero-load latency of an NoC with wormhole switching is [272]

$$T_{network} = hops \cdot t_r + t_c + \frac{L_p}{b}, \tag{9-1}$$

where the first term is the routing delay, t_c is the propagation delay along the wires of the communication channel, which is also called a buss here for simplicity, and the third term is the serialization delay of the packet. *Hops* is the average number of routers that a packet traverses to reach the destination node, t_r is the router delay, L_p is the length of the packet in bits, and b is the bandwidth of the communication channel defined as $b \equiv w_c f_c$, where w_c is the width of the channel in bits and f_c is the inverse of the propagation delay of a bit along the longest communication channel.

Since the number of planes that can be stacked in a 3-D NoC is constrained by the target technology, n_3 is also constrained. Furthermore, n_1, n_2, and n_3 are not necessarily equal. The average number of hops in a 3-D NoC is

$$hops = \frac{n_1 n_2 n_3 (n_1 + n_2 + n_3) - n_3 (n_1 + n_2) - n_1 n_2}{3(n_1 n_2 n_3 - 1)}, \tag{9-2}$$

assuming dimension-order routing where the minimum distance paths are used for routing the packets between any source-destination node pair. The number of hops in (9-2) can be divided into two components — the average number of hops within the two dimensions n_1 and n_2, and the average number of hops within the third dimension n_3,

$$hops_{2-D} = \frac{n_3 (n_1 + n_2)(n_1 n_2 - 1)}{3(n_1 n_2 n_3 - 1)}, \tag{9-3}$$

$$hops_{3-D} = \frac{(n_3^2 - 1) n_1 n_2}{3(n_1 n_2 n_3 - 1)}. \tag{9-4}$$

The delay of the router t_r is the sum of the delay of the arbitration logic t_a and the delay of the switch t_s, which in this chapter is considered to be implemented with a classic crossbar switch [272],

$$t_r = t_a + t_s. \tag{9-5}$$

The delay of the arbiter can be described from [273]

$$t_a = (21(1/4)\log_2 p + 14(1/12) + 9)\tau, \tag{9-6}$$

where p is the number of ports of the router and τ is the delay of a minimum-sized inverter for the target technology. Note that (9-6) exhibits a logarithmic dependence on the number of router ports. The length of the crossbar switch also depends on the number of router ports and the width of the buss,

$$l_s = 2(w_t + s_t)w_c p, \tag{9-7}$$

where w_t and s_t are, respectively, the width and spacing or, alternatively, the pitch of the interconnect and w_c is the width of the communication channel in bits. Consequently, the worst case delay of the crossbar switch is determined by the longest path within the switch, which is equal to (9-7).

The delay of the communication channel t_c is

$$t_c = t_v hops_{3-D} + t_h hops_{2-D}, \tag{9-8}$$

where t_v and t_h are the delay of the vertical and horizontal channels, respectively (see Figure 9-10b). Note that if $n_3 = 1$, (9-8) describes the propagation delay of a 2-D NoC. Substituting (9-8) and (9-5) into (9-1), the overall zero-load network latency for a 3-D NoC is

$$T_{network} = hops(t_a + t_s) + hops_{2-D} t_h + hops_{3-D} t_v + \frac{L_p}{w_c} t_h. \tag{9-9}$$

To characterize t_s, t_h, and t_v, the models described in [274] are adopted, where repeaters implemented as simple inverters are inserted along the interconnect. According to these models, the propagation delay and rise time of a single interconnect stage for a step input, respectively, are

$$t_{di} = 0.377\frac{r_i c_i l_i^2}{k_i^2} + \left(R_{d0}C_0 + \frac{R_{d0}c_i l_i}{h_i k_i} + \frac{r_i l_i C_{g0} h_i}{k_i}\right), \tag{9-10}$$

$$t_{ri} = 1.1\frac{r_i c_i l_i^2}{k_i^2} + 2.75\left(R_{r0}C_0 + \frac{R_{r0}c_i l_i}{h_i k_i} + \frac{r_i l_i C_{g0} h_i}{k_i}\right), \tag{9-11}$$

where r_i (c_i) is the per unit length resistance (capacitance) of the interconnect and l_i is the total length of the interconnect. The index i is used to notate the various interconnect delays included in the network (*i.e.*, $i \in \{s,v,h\}$). h_i and k_i denote the number and size of the repeaters, respectively, and C_{go} and C_0 represent the gate and total input capacitance of a minimum sized device, respectively. C_0 is the summation of the gate and drain capacitance of the device. R_{ro} and R_{do} describe the equivalent output resistance of a minimum-sized device for the propagation delay and transition time of a minimum-sized inverter, respectively, where the output resistance is approximated as

$$R_{r(d)0} = K_{r(d)} \frac{V_{dd}}{I_{dn0}}. \qquad (9\text{-}12)$$

K denotes a fitting coefficient, and I_{dn0} is the drain current of an NMOS device at both V_{ds} and V_{gs} equal to V_{dd}. The value of these device parameters are listed in Table 9-3. A 45-nm technology node is assumed, and SPICE simulations of the predictive technology library are used to determine the individual parameters [233], [234].

Table 9-3 Interconnect and Design Parameters, 45-nm Technology.

Parameter	Value	
	NMOS	PMOS
W_{min}	100 nm	250 nm
I_{dsat}/W	1115 µA/µm	349 µA/µm
V_{dsat}	478 mV	-731 mV
V_t	257 mV	-192 mV
α	1.04	1.33
I_{sub0}	48.8 nA	
I_{g0}	0.6 nA	
V_{dd}	1.1 Volts	
Temp.	110 °C	
K_d	0.98	
K_r	0.63	
C_{g0}	512 fF	
C_{d0}	487 fF	
τ	17 ps	

To include the effect of the input slew rate on the total delay of an interconnect stage, (9-10) and (9-11) are further refined by including an additional coefficient γ as in [275],

$$\gamma_r = \frac{1}{2} - \frac{1 - V_{tn}/V_{dd}}{1 - a_n}.$$

(9-13)

By substituting the subscript n with p, the corresponding value for a falling transition is obtained. The average value γ for γ_r and γ_f is used to describe the effect of the transition time on the interconnect delay. The overall interconnect delay can therefore be described as

$$t_i = k(t_{di} + \gamma t_{ri}) = a_1 \frac{r_i}{c_i} + a_2 \left(R_0 C_0 k + \frac{R_0 c_i l_i}{h} + R_i C_{g0} h \right),$$

(9-14)

where R_0, a_1, and a_2 are described in [276] and the index i denotes the various interconnect structures such as the crossbar switch ($i \equiv s$), the horizontal buss ($i \equiv h$), and the vertical buss ($i \equiv v$).

For minimum delay, the size h and number k of repeaters are determined by setting the partial derivative of t_i with respect to h_i and k_i, respectively, equal to zero and solving for h_i and k_i [23], [276],

$$k_i^* = \sqrt{\frac{a_1 r_i c_i l_i^2}{a_2 R_0 C_0}},$$

(9-15)

$$h_i^* = \sqrt{\frac{R_0 c_i l_i}{r_i C_{g0}}}.$$

(9-16)

The expression in (9-14) only considers *RC* interconnects. An *RC* model is sufficiently accurate to characterize the delay of a crossbar switch since the length of the longest wire within the crossbar switch and the signal frequencies are such that inductive behavior is not prominent [278]. For the buss lines, however, inductive behavior can appear. For this case, suitable expressions for the delay and repeater insertion characteristics can be adopted from [277]. For the target operating frequencies (1 to 2 GHz) and buss length (<2 mm) considered in this chapter, an *RC* interconnect model provides sufficient accuracy [278]. In addition, for the vertical buss, $k_v = 1$ and $h_v = 1$, meaning that no repeaters are inserted and minimum-sized drivers are utilized. Repeaters are not necessary due to the short length of the vertical buss. Driving a buss with minimum-sized inverters can affect the resulting minimum latency and power dissipation topology, as discussed in the following sections. Note that the latency expression includes the effect of the input slew rate.

Also, since a repeater insertion methodology for minimum latency is applied, any further reduction in latency is due to the network topology.

The length of the vertical communication channel for the 3-D NoC shown in Figure 9-10 is

$$l_v = \begin{cases} L_v, & \text{for 2DIC} - \text{3D NoC,} & (9\text{-}17a) \\ n_p\, L_v, & \text{for 3DIC} - \text{3D NoC,} & (9\text{-}17b) \\ 0, & \text{for 2DIC} - \text{2D NoC and 3DIC} - \text{2D NoC,} & (9\text{-}17c) \end{cases}$$

where L_v is the length of a silicon-through (interplane) via connecting two routers on adjacent physical planes. n_p is the number of physical planes used to integrate each PE. The length of the horizontal communication channel is assumed to be

$$l_h = \begin{cases} \sqrt{A_{PE}}, & \text{for 2DIC} - \text{2D NoC and 2DIC} - \text{3D NoC,} & (9\text{-}18a) \\ 1.12\sqrt{A_{PE}/n_p}, & \text{for 3DIC} - \text{2D NoC and 3DIC} - \text{3D NoC } (n_p > 1). & (9\text{-}18b) \end{cases}$$

where A_{PE} is the area of the processing element. The area of all of the PEs, and, consequently, the length of each horizontal channel are assumed to be equal. For those cases where the PE is implemented in multiple physical planes, a coefficient is included to consider the effect of the interplane vias on the reduction in the ideal wirelength due to utilization of the third dimension. The value of this coefficient ($= 1.12$) is based on the layout of a crossbar switch designed with the fully depleted silicon-on-insulator (FDSOI) 3-D technology from MIT Lincoln Laboratory (MITLL) [140]. The same coefficient is also assumed for the design of the PEs on more than one physical plane. In the following section, expressions for the power consumption of a network with delay constraints are presented.

9.3.3 Power Consumption in 3-D NoC

Power dissipation is a critical issue in three-dimensional circuits. Although the total power consumption of 3-D systems is expected to be lower than that of mainstream 2-D circuits (since the global interconnects are shorter [141]), the increased power density is a challenging issue for this novel design paradigm. Therefore, those 3-D NoC topologies that offer low-power characteristics are of significant interest.

The different power consumption components for interconnects with repeaters are briefly discussed in this section. A low-power design methodology with delay constraints for the interconnect in an NoC is adopted from [276]. An expression for the total power

consumption per bit of a packet transferred between a source destination node pair is used as the basis for characterizing the power
consumption of an NoC for the 3-D topologies.

The power consumption components of an interconnect line with
repeaters are:

(a) Dynamic power consumption is the dissipated power due to the
charge and discharge of the interconnect and input gate capacitance
during a signal transition, and can be described by

$$P_{di} = a_s f (c_i l_i + h_i k_i C_o) V_{dd}^2, \tag{9-19}$$

where f is the clock frequency and a_s is the switching factor [279]. A
value of 0.15 is assumed here; however, for NoC, the switching factor
can vary considerably. This variation, however, does not affect the
power comparison for the various topologies as the same switching
factor is incorporated in each term for the total power consumed
per bit of the network (the absolute value of the power consumption,
however, changes).

(b) Short-circuit power is due to the DC current path that exists in a
CMOS circuit during a signal transition when the input signal voltage
changes between V_{tn} and $V_{dd} + V_{tp}$. The power consumption due to this
current is described as short-circuit power and is modeled in [280] by

$$P_{si} = \frac{4 a_s f I_{do}^2 t_{ri}^2 V_{dd} k_i h_i^2}{V_{dsat} G C_{effi} + 2 H I_{do} t_{ri} h_i}, \tag{9-20}$$

where I_{do} is the average drain current of the NMOS and PMOS
devices operating in the saturation region and the value of the coefficients G and H are described in [281]. Due to resistive shielding of
the interconnect capacitance, an effective capacitance is used in (9-
20) rather than the total interconnect capacitance. This effective
capacitance is determined from the methodology described in
[282] and [283].

(c) Leakage power is comprised of two power components, the subthreshold and gate leakage currents. The subthreshold power consumption is due to current flowing during the cut-off region (below
threshold), causing I_{sub} current to flow. The gate leakage component
is due to current flowing through the gate oxide, denoted as I_g. The
total leakage power can be described as

$$P_{li} = h_i k_i V_{dd} (I_{sub0} + I_{g0}), \tag{9-21}$$

where the average subthreshold I_{sub0} and gate I_{g0} leakage current of
the NMOS and PMOS transistors is used in (9-21).

The total power consumption with delay constraint T_0 for a single line of a crossbar switch P_{stotal}, horizontal buss P_{htotal}, and vertical buss P_{vtotal} is, respectively,

$$P_{stotal}(T_0 - t_a) = P_{di} + P_{si} + P_{li.} \tag{9-22}$$

$$P_{htotal}(T_0) = P_{di} + P_{si} + P_{li.} \tag{9-23}$$

$$P_{vtotal}(T_0) = P_{di} + P_{si} + P_{li.} \tag{9-24}$$

The power consumption of the arbitration logic is not included in (9-22), since most of the power is consumed by the crossbar switch and the buss interconnect, as discussed in [284]. Note that for a crossbar switch, the additional delay t_a of the arbitration logic poses a stricter delay constraint on the power consumption of the switch, as shown in (9-22). The minimum power consumption with delay constraints is determined by the methodology described in [276], for which the optimum size h^*_{powi} and number k^*_{powi} of the repeaters for a single interconnect line is determined. Consequently, the minimum power consumption per bit between a source destination node pair in a NoC with a delay constraint is

$$P_{bit} = \text{hops}P_{stotal} + \text{hops}_{2-D}P_{htotal} + \text{hops}_{3-D}P_{vtotal.} \tag{9-25}$$

The effect of resistive shielding is also considered in determining the effective interconnect capacitance. Furthermore, since the repeater insertion methodology in [276] minimizes the power consumed by the repeater system, any additional decrease in power consumption is due only to the network topology. In the following section, those 3-D NoC topologies that exhibit the maximum performance and minimum power consumption with delay constraints are presented. Tradeoffs in determining these topologies are discussed, and the impact of the network parameters on the resulting optimum topologies is demonstrated for various network sizes.

9.3.4 **Performance and Power Analysis for 3-D NoC**

Several network parameters characterizing the topology of a network can significantly affect the speed and power of a system. The evaluation of these network parameters is discussed in Section 9.3.4.1. The improvement in network performance achieved by the 3-D NoC topologies is explored in Section 9.3.4.2. The distribution of nodes that produces the maximum performance is also discussed. The power consumption with delay constraints of a 3-D NoC and the topologies that yield the minimum power consumption of a 3-D NoC are presented in Section 9.3.4.3.

9.3.4.1 Parameters of 3-D Networks-on-Chip

The physical layer of a 3-D NoC consists of different interconnect structures, such as a crossbar switch, the horizontal buss connecting neighboring nodes on the same physical plane, and the vertical buss connecting nodes on different, not necessarily adjacent, physical planes. The device parameters characterizing the receiver, driver, and repeaters are listed in Table 9-3. The interconnect parameters reported in Table 9-4 are different for each type of interconnect within a network.

A typical interconnect structure is shown in Figure 9-11, where three parallel metal lines are sandwiched between two ground planes. This interconnect structure is considered for the crossbar switch (at the network nodes) where the intermediate metal layers are assumed to be utilized. The horizontal buss is implemented on the global metal layers, and, therefore, only the lower ground plane is present in this structure for a 2-D NoC. For a 3-D NoC, however, the substrate (front-to-back plane bonding) or a global metal layer of an upper plane (front-to-front plane bonding) behaves as a second ground plane. To incorporate this additional ground plane, the horizontal

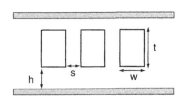

■ **FIGURE 9-11** Typical interconnect structure for intermediate metal layers.

Table 9-4 Interconnect parameters.

Interconnect Structure	Parameter	
	Electrical	**Physical**
Crossbar switch	$\rho = 3.07\ \mu\Omega\text{-cm}$	$w = 200$ nm
	$k_{ILD} = 2.7$	$s = 200$ nm
	$r_s = 614\ \Omega/\text{mm}$	$t = 250$ nm
	$c_s = 157.6$ fF/mm	$h = 500$ nm
Horizontal buss	$\rho = 2.53\ \mu\Omega\text{-cm}$	$w = 500$ nm
	$k_{ILD} = 2.7$	$s = 250\ (500)$ nm
	$r_h = 46\ \Omega/\text{mm}$	$t = 1100$ nm
	$c_h = 332.6\ (192.5)$ fF/mm	$h = 800$ nm
	$a_{3\text{-}D} = 1.02\ (1.06)$	-
Vertical buss	$\rho = 5.65\ \mu\Omega\text{-cm}$	$w = 1050$ nm
	$r_v = 51.2\ \Omega/\text{mm}$	$L_v = 10\ \mu\text{m}$
	$c_v = 600$ fF/mm	-

buss capacitance is changed by the coefficient $a_{3\text{-}D}$. A second ground plane decreases the coupling capacitance to an adjacent line, while the line-to-ground capacitance increases. The vertical buss is different from the other structures in that this buss is implemented by through silicon vias. These interplane vias can exhibit significantly different impedance characteristics as compared to traditional horizontal interconnect structures, as discussed in [233] and also verified by extracted impedance parameters. The electrical interconnect parameters are extracted using a commercial impedance extraction tool [224], while the physical parameters are extrapolated from the predictive technology library [233], [234] and the 3-D integration technology developed by MITLL for a 45-nm technology node [140]. The physical and electrical interconnect parameters are listed in Table 9-4. For each of the interconnect structures, a buss width of 64 bits is assumed. In addition, n_3 and n_p are constrained by the maximum number of physical planes n_{max} that can be vertically stacked. A maximum of eight planes is assumed. The constraints that apply for each of the 3-D NoC topologies shown in Figure 9-10 are

$$n_3 \leq n_{\max}, \quad \text{for 2D IC} - \text{3D NoC}, \tag{9-26a}$$

$$n_p \leq n_{\max}, \quad \text{for 3D IC} - \text{2D NoC}, \tag{9-26b}$$

$$n_3 n_p \leq n_{\max}, \quad \text{for 3D IC} - \text{3D NoC}. \tag{9-26c}$$

A small set of parameters is used as variables to explore the performance and power consumption of the 3-D NoC topologies. This set includes the network size or, equivalently, the number of nodes within the network N, the area of each processing element A_{PE}, which is directly related to the buss length as described in (9-18), and the maximum allowed interconnect delay when evaluating the minimum power consumption with delay constraints. The range of values for these variables is listed in Table 9-5. Depending on the network size, NoCs are roughly divided as small ($N = 16$ to 64 nodes), medium

Table 9-5 Network parameters.

Parameter	Values
N	16, 32, 64, 128, 256, 512, 1024, 2048
A_{PE} [mm^2]	0.5, 0.64, 0.81, 1.00, 1.56, 2.25, 4.00
T_0 [ps]	1000, 500

($N = 128$ to 256 nodes), and large ($N = 512$ to 2048 nodes) net-works. For multiprocessor SoC networks, sizes of up to $N = 256$ are expected to be feasible in the near future [268], [285], whereas for NoC with a finer granularity, where each PE corresponds to hardware blocks of approximately 100K gates, network sizes over a few thousands nodes are predicted at the 45-nm technology node [286]. Note that this classification of the networks is not strict and is only intended to facilitate the discussion in the following sections.

9.3.4.2 Performance Tradeoffs for 3-D NoC

The performance enhancements that can be achieved in NoC by utilizing the third dimension are investigated in this subsection. Each of the 3-D topologies decreases the zero-latency of the network by reducing different delay components, as described in (9-9). In addi-tion, the distribution of network nodes in each physical dimension that yields the minimum zero-load latency is shown to significantly change with the network and interconnect parameters.

2-D IC–3-D NoC

Utilizing the third dimension to implement an NoC results directly in a decrease in the average number of hops for packet switching. The average number of hops on the same plane $hops_{2-D}$ (the intraplane hops) and the average number of hops in the third dimension $hops_{3-D}$ (the interplane hops) are also reduced. Interestingly, the dis-tribution of nodes n_1, n_2, and n_3 that yields the minimum total num-ber of hops is not always the same as the distribution that minimizes the number of intraplane hops. This situation occurs particularly for small and medium networks, while for large networks, the distribu-tion of n_1, n_2, and n_3 that minimizes the $hops$ also minimizes $hops_{2-D}$.

In a 3-D NoC, the number of router ports increases from five to seven, increasing, in turn, both the switch and arbiter delay. Furthermore, a short vertical buss generally exhibits a lower delay than that of a rela-tively long horizontal buss. In Figure 9-12, the zero-load latency of the 2-D IC – 3-D NoC is compared to that of the 2-D IC–2-D NoC for different network sizes. A decrease in latency of 15.7% and 20.1% can be observed for $N = 128$ and $N = 256$ nodes, respectively, with $A_{PE} = 0.81$ mm^2.

The node distribution that produces the lowest latency varies with net-work size. For example, $n_{3max} = 8$ is not necessarily the optimum for small and medium networks, although by increasing n_3, more hops

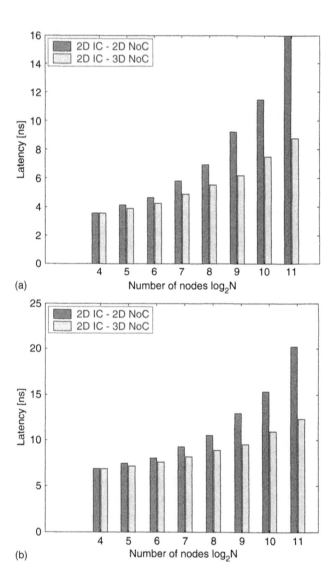

■ **FIGURE 9-12** Zero-load latency for various network sizes. (a) $A_{PE} = 0.81$ mm^2 and $c_h = 332.6$ fF/mm, (b) $A_{PE} = 4$ mm^2 and $c_h = 332.6$ fF/mm.

occur through the short, low-latency vertical channel. This result can be explained by considering the reduction in the number of hops that originates from utilizing the third dimension for packet switching. For small and medium networks, the decrease in the number of hops is small and cannot compensate the increase in the routing delay due to the increase in the number of ports of a router in a 3-D NoC. As the horizontal buss length becomes longer, however (*e.g.*, approaching 2 mm), $n_3 > 1$, and a slight decrease in the number of hops significantly decreases the overall delay, despite the increase in the routing delay for a 3-D NoC. As an example, consider a network with $log_2N = 4$ and $A_{PE} = 0.81$ mm^2. The minimum latency node distribution is $n_1 = n_2 = 4$ and $n_3 = 1$ (identical to a 2-D IC – 2-D NoC, as shown in Figure 9-12), while for $A_{PE} = 4$ mm^2, $n_1 = n_2 = 2$ and $n_3 = 4$.

The optimum node distribution can also be affected by the delay of the vertical channel. The repeater insertion methodology for minimum delay as described in Section 9.3.2 can significantly reduce the delay of the horizontal buss by inserting large-sized repeaters (*i.e.*, $h > 300$). In this case, the delay of the vertical buss becomes comparable to that of the horizontal buss with repeaters. Consider a network with $N = 128$ nodes. Two different node distributions yield the minimum average number of hops, specifically, $n_1 = 4$, $n_2 = 4$, and $n_3 = 8$ and $n_1 = 8$, $n_2 = 4$, and $n_3 = 4$. The first of the two distributions also results in the minimum number of intraplane hops$_{2-D}$, thereby reducing the latency component for the horizontal buss as described by (9.9). Simulation results, however, indicate that this distribution is not the minimum latency node distribution, as the delay due to the vertical channel is nonnegligible. For this reason, the latter distribution with $n_3 = 4$ is preferable, since a smaller number of hops$_{3-D}$ occurs, resulting in the minimum network latency.

3-D IC – 2-D NoC

For this type of three-dimensional network, the PEs are allowed to span multiple physical planes while the network effectively remains two-dimensional (*i.e.*, $n_3 = 1$). Consequently, the network latency is only reduced by decreasing the length of the horizontal buss, as described in (9-18). The routing delay component remains constant with this 3-D topology. Decreasing the horizontal buss length lowers both the communication channel delay and the serialization delay. In Figure 9-13, the decrease in latency that can be achieved by a 3-D IC – 2-D NoC is illustrated. A latency decrease of 30.2% and 26.4% can be observed for $N = 128$ and $N = 256$ nodes, respectively, with $A_{PE} = 2.25$ mm^2. The use of multiple physical planes reduces the latency; therefore, the optimum value for $n_p = n_{max}$ regardless of the network size and buss length.

In Figures 9-14a and 9-14b, the improvement in the network latency over a 2-D IC–2-D NoC for various network sizes and for different PE areas (*i.e.*, different horizontal buss length) is illustrated for the 2-D IC – 3-D NoC and 3-D IC – 2-D NoC topologies, respectively. Note that for the 2-D IC – 3-D NoC topology, the improvement in delay

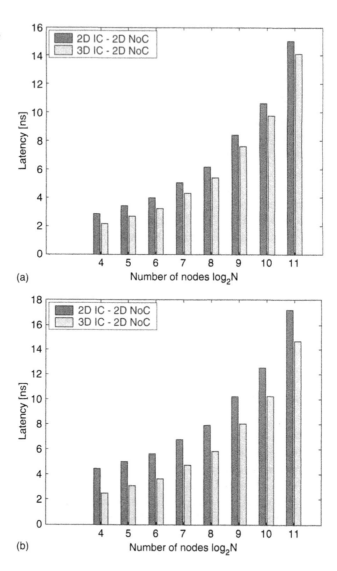

■**FIGURE 9-13** Zero-load latency for various network sizes. (a) $A_{PE} = 0.64$ mm^2 and $c_h = 192.5$ fF/mm, (b) $A_{PE} = 2.25$ mm^2 and $c_h = 192.5$ fF/mm.

is smaller for PEs with a larger area or, equivalently, with longer buss lengths independent of the network size. For longer buss lengths, the buss latency comprises a larger portion of the total network latency. Since for a 2-D IC –3-D NoC only the hop count is reduced, the improvement in latency is lower for longer buss lengths. Alternatively, the improvement in latency is greater for PEs with a larger area independent of the network size for 3-D IC – 2-D NoC. This situation is due to the significant reduction in the PE area (or buss length) that

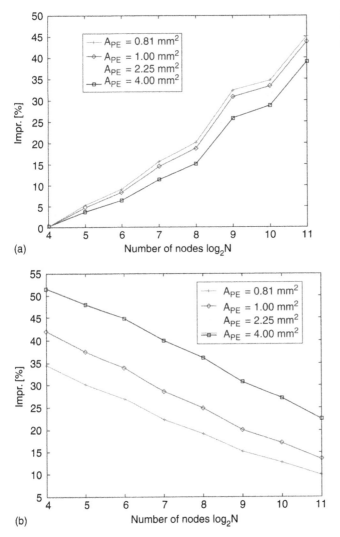

is achieved with this topology. Consequently, there is a tradeoff in the latency of a NoC that depends on both the network size and the area of the PEs. In Figure 9-14a, the improvement is not significant for small networks (all of the curves converge to approximately zero) in 2-D IC–3-D NoC, while this situation does not occur for 3-D IC – 2-D NoC. This behavior is due to the increase in the delay of the network router as the number of ports increases from five to seven for 2-D IC – 3-D NoC, which is a considerable portion of the network latency for small networks. Note that for 3-D IC – 2-D NoC, the network essentially remains two dimensional, and therefore the delay of the router for this topology does not increase. To achieve the minimum delay, a 3-D NoC topology that exploits these tradeoffs is described in the following subsection.

3-D IC – 3-D NoC

This topology offers the greatest decrease in latency over the aforementioned three-dimensional topologies. The 2-D IC – 3-D NoC topology decreases the number of hops, while the buss and serialization delays remain constant. With the 3-D IC – 2-D NoC, the buss and serialization delay is smaller, but the number of hops remains unchanged. With the 3-D IC – 3-D NoC, all of the latency components can be decreased by assigning a portion of the available physical planes for implementing the network, while the remaining planes of the stack are used for the PE. The resulting decrease in network latency as compared to a standard 2-D IC – 2-D NoC and the other two 3-D topologies is illustrated in Figure 9-15. A decrease in latency of 40% and 36% can be observed for $N = 128$ and $N = 256$ nodes, respectively, with $A_{PE} = 4$ mm^2. Note that the 3-D IC – 3-D NoC topology achieves the greatest savings in latency by optimally balancing n_3 with n_p.

For certain network sizes, the performance of the 3-D IC – 2-D NoC is identical to either the 2-D IC – 3-D NoC or 3-D IC – 2-D NoC. This behavior occurs because for large network sizes, the delay due to the large number of hops dominates the total delay, and, therefore, the latency can be primarily reduced by decreasing the average number of hops ($n_3 = n_{max}$). For small networks, the buss delay is large, and the latency savings is typically achieved by reducing the buss length ($n_p = n_{max}$). For medium networks, though, the optimum topology is obtained by dividing n_{max} between n_3 and n_p to satisfy (9-26c). This distribution of n_3 and n_p as a function of the network size and buss length is illustrated in Figure 9-16.

Note the shift in the value of n_3 and n_p as the PE area A_{PE} or, equivalently, the buss length increases. For long busses, the delay of the communication channel becomes dominant, and therefore the smaller number of hops for medium-sized networks cannot significantly decrease the total delay. Alternatively, further decreasing the buss length by implementing the PEs in a greater number of physical planes leads to a larger savings in delay.

The suggested optimum topologies for various network sizes (namely, small, medium, and large networks) also depend on the

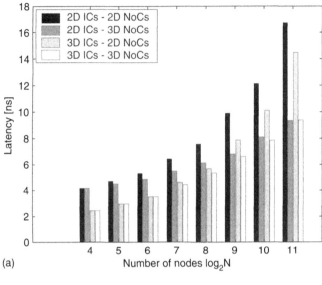

(a)

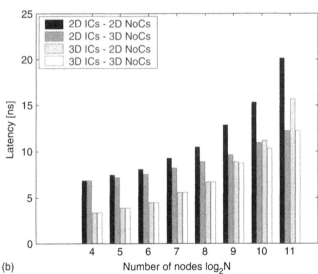

(b)

■ **FIGURE 9-15** Zero-load latency for various network sizes. (a) $A_{PE} = 1$ mm^2 and $c_h = 332.6$ fF/mm, (b) $A_{PE} = 4$ mm^2 and $c_h = 332.6$ fF/mm.

interconnect parameters of the network. Consequently, a change in the optimum topology for different network sizes can occur when different interconnect parameters are considered. Despite the sensitivity of the topologies to the interconnect parameters, the tradeoff between the number of hops and the buss length for various 3-D topologies (see Figures 9-14 and 9-16) can be exploited to improve the performance of an NoC. In the following subsection,

■ **FIGURE 9-16** n_3 and n_p values for minimum zero-load latency for various network sizes. (a) $A_{PE} = 1$ mm^2 and $c_h = 332.6$ fF/mm, (b) $A_{PE} = 4$ mm^2 and $c_h = 332.6$ fF/mm.

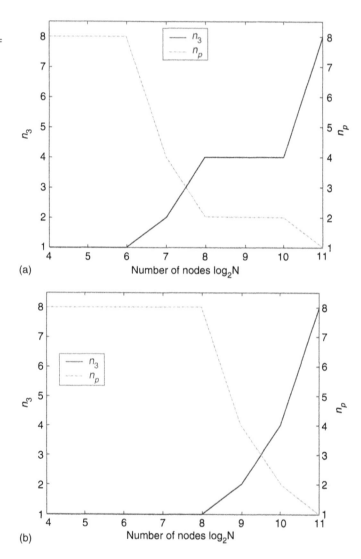

the topology that yields the minimum power consumption with delay constraints is described. The distribution of nodes for that topology is also discussed.

9.3.4.3 Power Consumption in 3-D NoC

The various power consumption components for the interconnect within an NoC are analyzed in Section 9.3.3. The methodology presented in [276] is applied here to minimize the power consumed by these interconnects while satisfying the specified operating frequency of the network. Since a power minimization methodology is applied to the buss lines, the power consumed by the network can only be further reduced by the choice of the network topology. In addition, the power consumption also depends on the target operating frequency, as discussed later in this section.

As with the zero-load latency, each topology affects the power consumption of the network in a different way. From (9.25), the power consumption can be reduced by either decreasing the number of hops for the packet or decreasing the buss length. Note that by reducing the buss length, the interconnect capacitance is not only reduced but also the number and size of the repeaters required to drive the lines are decreased, resulting in a greater savings in power. The effect of each of the 3-D topologies on the power consumption of an NoC is investigated in this section.

2-D IC – 3-D NoC
Similar to the network latency, the power consumption is decreased in this topology by reducing the number of hops for packet switching. Again, the increase in the number of ports is significant; however, the impact from this increase is not as important as that on the latency of the network. A three-dimensional network, therefore, can reduce power even in small networks. The power savings achieved with this topology is depicted in Figure 9-17 for various network sizes, where the savings is greater in larger networks. This situation occurs because the reduction in the average number of hops for a three-dimensional network increases for larger network sizes. A power savings of 26.1% and 37.9% is achieved for $N = 128$ and $N = 512$, respectively, with $A_{PE} = 1 \text{ mm}^2$.

3-D IC – 2-D NoC
With this topology, the number of hops in the network is the same as for a two-dimensional network. The horizontal buss length, however, is shorter by implementing the PEs in more than one physical plane. The greater the number of physical planes that can be integrated in a 3-D system, the larger the power savings, meaning that the optimum value for n_p with this topology is always n_{max} regardless of the network size and operating frequency. The achieved savings is practically limited by the number of physical planes that can be integrated in a 3-D technology. The power savings for various network sizes are shown in Figure 9-18. Note that for this type of NoC, the maximum performance topology is identical to the minimum power consumption topology, as the key element of both objectives originates solely from

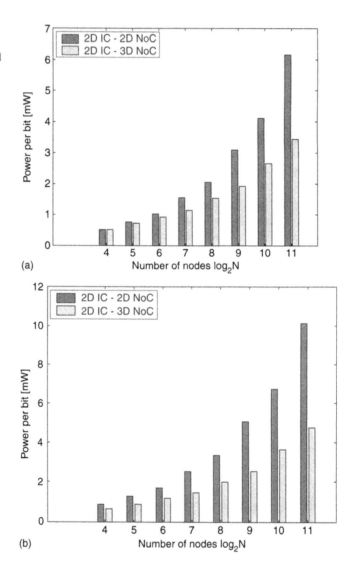

■ FIGURE 9-17 Power consumption with delay constraints for various network sizes. (a) $A_{PE} = 1$ mm^2, $c_h = 332.6$ fF/mm, and $T_0 = 500$ ps, (b) $A_{PE} = 4$ mm^2, $c_h = 332.6$ fF/mm, and $T_0 = 500$ ps.

the shorter buss length. The savings in power is approximately 35% when $A_{PE} = 0.64$ mm^2 for every network size as the per cent reduction in the buss length is the same for each network size.

3-D IC – 3-D NoC

Allowing the available physical planes to be utilized either for the third dimension of the network or for the PEs, the 3-D IC – 3-D NoC scheme achieves the greatest savings in power in addition to the minimum

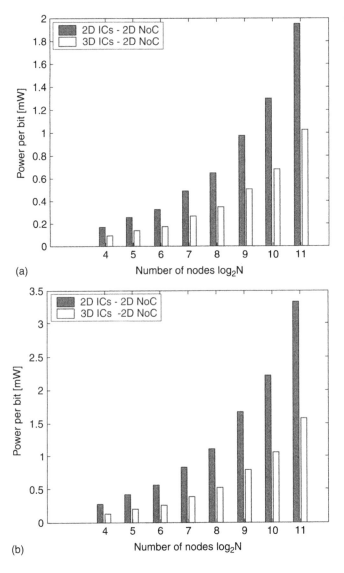

(a)

(b)

■ **FIGURE 9-18** Power consumption with delay constraints for various network sizes. (a) $A_{PE} = 0.64$ mm^2, $c_h = 192.5$ fF/mm, and $T_0 = 1000$ ps, (b) $A_{PE} = 2.25$ mm^2, $c_h = 192.5$ fF/mm, and $T_0 = 1000$ ps.

■ **FIGURE 9-18** Power consumption with delay constraints for various network sizes. (a) $A_{PE} = 0.64$ mm^2, $c_h = 192.5$ fF/mm, and $T_0 = 1000$ ps, (b) $A_{PE} = 2.25$ mm^2, $c_h = 192.5$ fF/mm, and $T_0 = 1000$ ps.

delay, as discussed in the previous subsection. The distribution of nodes along the physical dimensions, however, that produces either the minimum latency or the minimum power consumption for every network size is not necessarily the same. This nonequivalence is due to the different degree of importance of the average number of hops and the buss length in determining the latency and power consumption of a network. In Figure 9-19, the power consumption of the 3-D IC – 3-D NoC topology is compared to the three-dimensional topologies

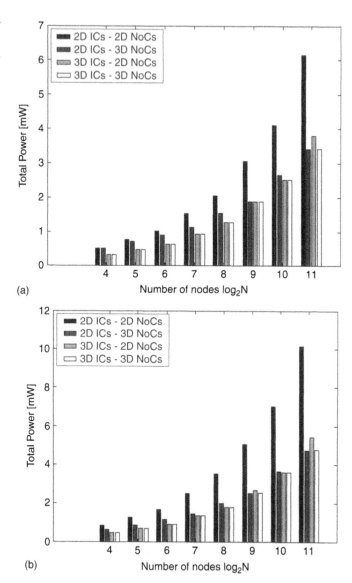

■ **FIGURE 9-19** Power consumption with delay constraints for various network sizes. (a) $A_{PE} = 1$ mm^2, $c_h = 332.6$ fF/mm, and $T_0 = 500$ ps, and (b) $A_{PE} = 4$ mm^2, $c_h = 332.6$ fF/mm, and $T_0 = 500$ ps.

previously discussed. A power savings of 38.4% is achieved for $N = 128$ with $A_{PE} = 1$ mm^2. For certain network sizes, the power consumption of the 3-D IC – 3-D NoC topology is the same as that of the 2-D IC – 3-D NoC and 3-D IC – 2-D NoC topologies. For the 2-D IC – 3-D NoC, the power consumption is primarily decreased by reducing the number of hops for packet switching, while for the 3-D IC – 2-D NoC, the NoC power dissipation is decreased by shortening the buss length. The former approach typically benefits small networks, while the latter approach yields lower power consumption for large networks. For medium-sized networks and depending on the network and intercon- nect parameters, nonextreme values for the n_3 and n_p parameters (*e.g.*, $1 < n_3 < n_{max}$ and $1 < n_p < n_{max}$) are required to produce the minimum power consumption topology.

Note that this work emphasizes the latency and power consumption of a network, neglecting the performance requirements of the individual PEs. If the performance of the individual PEs is important, only one 3-D topology may be available; however, even with this constraint, a sig- nificant savings in latency and power can be achieved since in almost every case the network latency and power consumption can be decreased as compared to a 2-D IC – 2-D NoC topology. Furthermore, as previously mentioned, if the available topology is the 2-D IC – 3-D NoC, setting n_3 equal to n_{max} is not necessarily the optimum choice.

The zero-load network latency and power consumption expressions capture the effect of the topology; yet these models do not incorporate the effects of the routing scheme and traffic load. Alternatively, these models can be treated as lower bounds for both the latency and the power consumption of the network. Since minimum distance paths and no contention are implicitly assumed in these expressions, nonmi- nimal path-routing schemes and heavy traffic loads will result in increasing both the latency and power consumption of the network. The routing algorithm is determined at the upper layers, other than the physical layer, comprising the communication protocol implemen- ted by the network. In addition, the traffic patterns depend upon the application being executed by the network. The effect of each of the parameters on the performance of 3-D NoCs is explored in the follow- ing section by utilizing a novel network simulator.

9.3.5 **Design Aids for 3-D NoCs**[*]

To evaluate the performance of emerging 3-D topologies for different applications, effective design aids are required. A recently developed network simulator that evaluates the effectiveness of different 3-D topologies is described in the following subsection, 9.3.5.1. Simula- tions of a broad variety of 3-D mesh- and torus-based topologies as well as traffic patterns are discussed in subsection 9.3.5.2.

---

[*]This section, 9.3.5, was contributed by Professor Dimitrios Soudris of the Democritus University of Thrace.

9.3.5.1 3-D NoC Simulator

The core of the 3-D NoC simulator is based on the *Worm_Sim* NoC simulator [287], which has been extended to support 3-D topologies. In addition to 3-D meshes and tori, variants of these topologies are supported. Related routing schemes have also been adapted to provide routing in the vertical direction. An overview of the characteristics and capabilities of the simulator is depicted in Figure 9-20.

Several fundamental characteristics of a network topology are reported, including the energy consumption, average packet latency, and router area. The reported energy consumption includes the energy consumed by each component of the network, such as the crossbar switch and other circuitry within the network router, and the interconnect busses (*i.e.*, link energy). Consequently, this simulator is a useful tool for exploring early decisions related to the system architecture.

An important capability of this tool is that variations of a basic 3-D mesh and torus topology can be efficiently explored. These topologies are characterized by heterogeneous interconnectivity, combining 2-D and 3-D network routers within the same network. The primary difference between these routers is that since a 3-D network router is connected to network routers on adjacent planes, a 3-D router has

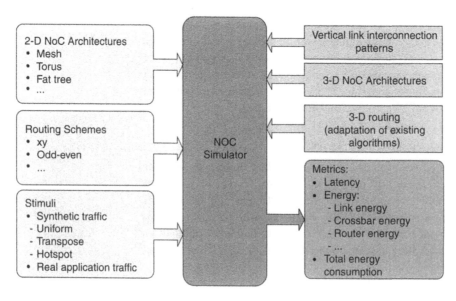

■ **FIGURE 9-20** An overview of the 3-D NoC simulator.

two more ports than a 2-D network router. Consequently, a 3-D network router consumes a larger area and dissipates more power, yet
provides greater interconnectivity.

Although these topologies typically have a greater delay than a straightforward 3-D topology, the savings in energy can be significant; in particular, those applications where speed is not the primary objective.
Another application area that can benefit from these topologies is those
applications where the data packets propagate over a small number of
routers within the network. Alternatively, the spatial distribution of the
required hops to propagate a data packet within the on-chip network
is small. Since different applications can produce diverse types of traffic,
an efficient traffic model is required. The simulator includes the traffic
model used within the Trident tool [288]. This model consists of several
parameters that characterize the spatial and temporal traffic across a network. The temporal parameters include the number of packets and
rate at which these packets are injected into a router. The spatial parameters include the distance that a packet travels within the network
and the portion of the total traffic injected by each router into the
network [288].

To generate these heterogeneous mesh and torus topologies, a distribution of the 2-D and 3-D routers needs to be determined. The
following combinations of 2-D and 3-D routers for a 3-D mesh and
torus are considered [289]:

Uniform: The 3-D routers are uniformly distributed over the different
planes. In this scheme, the 3-D routers are placed on each physical
plane of the 3-D NoC. If no 2-D routers are inserted within the
network, the topology is a 3-D mesh or torus (see Figures 9-10 and 9-
21a). In the case where both 2-D and 3-D routers are used within the
network, the location of the routers on each plane is determined as
follows:

- Place the first 3-D router on the (X, Y, Z) position of each
  plane.

- The four neighboring 2-D routers are placed in positions $(X + r + 1,
  Y, Z)$, $(X - r - 1, Y, Z)$, $(X, Y + r + 1, Z)$ and $(X, Y - r - 1, Z)$. The
  parameter r represents the periodicity (or frequency) of the 2-D routers within each plane. Consequently, a 2-D router is inserted every r
  3-D routers in each direction within a plane. This scheme is exempli-
  fied in Figure 9-21b, where one plane of a 3-D NoC with $r = 3$
  is shown.

Center: The 3-D routers are located at the center of each plane, as shown in Figure 9-21c. Since 3-D routers only exist at the center of a plane, only 2-D routers are placed along the outer region of each plane to connect the neighboring network nodes within the same plane.

Periphery: The 3-D routers are located at the periphery of each plane (see Figure 9-21d). This combination is complementary to the **Center** placement.

Full Custom: The position of the 3-D routers is fully customized based on the requirements of the application, while minimizing the area occupied by the routers, since fewer 3-D routers are used. This configuration, however, produces an irregular design that does not adapt well to changes in the network functionality and application.

As illustrated in Figure 9-20, different routing schemes are supported by the simulator. These algorithms are extended from the *Worm_Sim* simulator to support routing in the third dimension:

■ **FIGURE 9-21** Position of the vertical interconnection links for each plane within a 3-D NoC (each plane is a 6 × 6 mesh): (a) fully connected 3-D NoC, (b) uniform distribution of vertical links, (c) vertical links at the center of the NoC, and (d) vertical links at the periphery of the NoC.

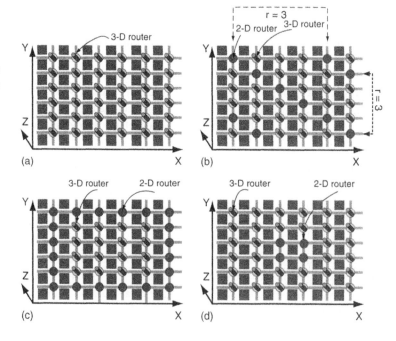

XYZ-OLD, which is an extended version of the shortest path XY routing algorithm.

XYZ, which is based on XY routing where this algorithm determines which direction produces a lower delay, forwarding the packet in this direction.

ODD-EVEN is the odd-even routing scheme as presented in [209]. In this scheme, the packets can take turns to avoid deadlock situations.

Consequently, the performance of a 3-D NoC can be evaluated under a broad and diverse set of parameters that reinforce the exploratory capabilities of the tool. In general, the following parameters of the simulator are configured:

- The NoC architecture, which can be a two- or three-dimensional mesh or torus, where the dimensions of the network in the x-, y-, and z-directions are n_1, n_2, and n_3, respectively.

- The type of input traffic and traffic load.

- The routing algorithm.

- The vertical link configuration file that defines whether a vertical link (required in 3-D routers) is present.

- The router model and related energy and delay models.

In the following section, several 3-D NoC topologies are explored under different traffic patterns and loads.

9.3.5.2 Evaluation of 3-D NoCs under Different Traffic Scenarios

To compare the performance of 3-D topologies with conventional 2-D meshes and tori, two different network sizes with 64 and 144 nodes are considered. Each of these case studies is implemented both in two and three dimensions. In two dimensions, the network nodes are connected to form 8×8 and 12×12 2-D meshes and tori, respectively. Alternatively, the three-dimensional topologies of the target networks are implemented on four physical planes (*i.e.*, $n_3 = 4$). Consequently, the dimensions of the two networks ($n_1 \times n_2 \times n_3$) are $4 \times 4 \times 4$ and $6 \times 6 \times 4$, respectively. Metrics for comparing the 2-D and 3-D topologies are the average packet latency, dissipated energy, and physical area. The area of the processing elements is excluded since this area is the same in both the 2-D and 3-D topologies. Topologies that contain processing elements on multiple planes (see Figure 9-10c) are not considered in this analysis.

Although the 3-D topologies exhibit superior performance as compared to the 2-D meshes and tori, a topology that combines 2-D and 3-D routers can be beneficial for certain traffic patterns and loads. Using a smaller number of 3-D routers in a network results in a smaller area and possibly lower power. This situation is due to the smaller number of ports required by a 2-D router, which has two fewer ports as compared to a 3-D router. Consequently, several combinations of 2-D and 3-D routers within a 3-D topology have been evaluated, some of which are illustrated in Figure 9-21. Note that for each combination of 2-D and 3-D routers, the location of these routers within a 3-D on-chip network is the same on each plane of the network. In the case studies, ten different combinations of 2-D and 3-D routers are compared in terms of energy, delay, and area. These combinations are described below, where the number (and per cent in parentheses) of 2-D and 3-D routers within a $4 \times 4 \times 4$ NoC is provided for the sake of clarity:

- **Full:** All of the processing elements are connected to three-dimensional routers (number of 3-D routers: 64 (100%)). This combination corresponds to a fully connected 3-D mesh (see Figure 9-10a) or torus network.

- **Uniform-based:** 2-D routers are connected to specific processing elements within each plane of the 3-D network. The distribution of the 2-D routers is controlled by the parameter r, as discussed in the previous section. The chosen values are three (by_three), four (by_four), and five (by_five). The corresponding number of 3-D routers is 44 (68.75%), 48 (75%), and 52 (81.25%). These combinations contain a decreasing number of 2-D routers, approaching a fully connected 3-D mesh or torus.

- **Odd:** In this combination, all of the routers within a row are of the same type (i.e., either 2-D or 3-D). The type of router alternates among rows (the number of 3-D routers is 32 (50%)).

- **Edges:** A portion of the processing elements is located at the center of each plane and connected to 2-D routers with dimensions $n_{x(2-D)} \times n_{x(2-D)}$, while the remaining network nodes are connected to 3-D routers. For the example network, $n_{x(2-D)} = 2$ and, consequently, the number of 3-D routers is 48 (75%).

- **Center:** A segment of the processing elements located at the center of each plane is connected to 3-D routers with dimensions $n_{x(3-D)} \times$

$n_{x(3-D)}$ while the remaining PEs are connected to 2-D routers. For the example network, the number of 3-D routers is 16 (25%).

- **Side-based:** The processing elements along a side (*e.g.*, an outer row) of each plane are connected to 2-D routers. The combinations have the PEs along one (*one_side*), two (*two_side*), or three (*three_side*) sides connected to 2-D routers. Consequently, the number of 3-D routers for each pattern is 48 (75%), 36 (56.25%), and 24 (37.5%), respectively. These combinations have an increasing number of 2-D routers, approaching the same number of routers as in a 2-D mesh or torus.

The interconnectivity for the two network sizes is evaluated under different traffic patterns and loads. The parameters of the traffic model used in the 3-D NoC simulator have been adjusted to produce the following common types of traffic patterns [291]:

Uniform: The traffic is uniformly distributed across the network, with the network nodes receiving approximately the same number of packets.

Transpose: In this traffic scheme, packets originating from a node at location (a, b, c) reach the node at the destination $(n_1 - a, n_2 - b, n_3 - c)$, where n_1, n_2, n_3 are the dimensions of the network.

Hot spot: A small number of network nodes (*i.e.*, hot spot nodes) receive an increasing number of packets as compared to the majority of the nodes, which is modeled as uniformly receiving packets. The hot spot nodes within a 2-D NoC are positioned in the middle of each quadrant of the network. Alternatively, in a 3-D NoC, a hot spot is located in the middle of each plane.

The traffic loads are low, normal, and high. The heavy load has a 50% increase in traffic, whereas the low load has a 90% decrease in traffic as compared to a normal load.

The energy consumption in Joules and the average packet latency in cycles are compared for each of these patterns. The energy model of the NoC simulator is an architectural level model that determines the energy consumed by propagating a single bit across the network [292]. This model includes the energy of a bit through the various components of the network, such as the buffer and switch within a router and the interconnect buss between two neighboring routers. Note that only the dynamic component of the consumed energy is considered in this model [292]. In addition, for each topology and combination of 2-D and 3-D routers, the total area of the routers is determined

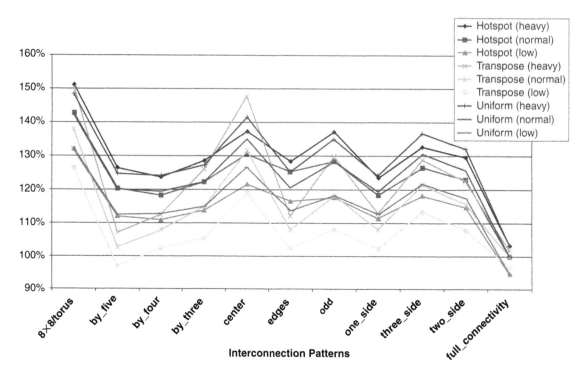

■FIGURE 9-22 Effect of traffic load on the latency of a 2-D and 3-D torus NoC for each type of traffic and **XYZ** routing.

based on the gate equivalent of the switching fabric [293]. The objective is to determine which of these combinations results in higher network performance as compared to the 2-D and fully connected 3-D NoC. All of the simulations are performed for 200,000 cycles.

The average packet latency of the differently interconnected NoCs for a torus architecture is depicted in Figure 9-22. The latency is normalized to the average packet latency of a fully connected 3-D NoC under normal load conditions and for each traffic scheme. As expected, the network latency increases proportionally with traffic load.

Mesh topologies exhibit similar behavior, though the latency is higher due to the decreased connectivity as compared to the torus topologies. This behavior is depicted in Figure 9-23, where the latency of a 64-node mesh and torus NoC are compared (the basis for the latency normalization is the average packet latency of a fully connected 3-D torus). The mesh topologies exhibit an increased packet latency of 34% as compared to the torus topology for the same traffic pattern, traffic load, and routing algorithm.

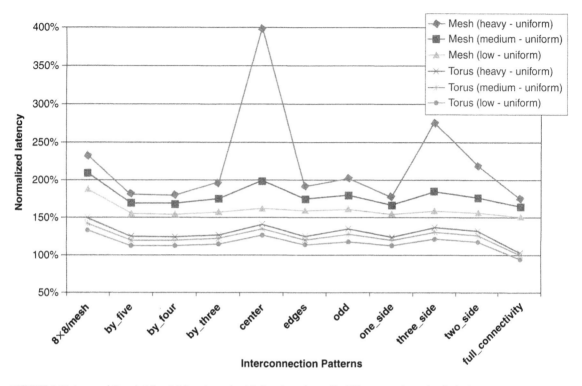

■ **FIGURE 9-23** Latency of 64-node 2-D and 3-D meshes and tori NoCs under uniform traffic, XYZ routing, and several traffic loads.

The results of employing partial vertical connectivity (*i.e.*, a combination of 2-D and 3-D routers) within a 3-D mesh network with uniform traffic, medium traffic load, and **XYZ-OLD** routing are illustrated in Figure 9-24. The energy consumption, average packet latency, router area, and per cent of 2-D routers for the $4 \times 4 \times 4$ and $6 \times 6 \times 4$ mesh architectures are illustrated in Figures 9-24a and 9-24b, respectively. All of these metrics are normalized to a fully connected 3-D NoC.

The advantages of a 3-D NoC as compared to a 2-D NoC are depicted in Figure 9-24a. In this case, the 8×8 mesh dissipates 39% more energy and exhibits a 29% higher packet delivery latency as compared to a fully connected 3-D NoC. The overall area of the routers, however, is 71% of the area of a fully connected 3-D NoC, since all of the routers are two-dimensional. Employing the *by_five* combination results in a 3% reduction in energy and 5% increase in latency. In this combination, only 81% of the routers are 3-D, resulting in 5% smaller area for the switching logic. For a larger network (see Figure 9-24b), several router combinations are superior to a fully connected 3-D NoC. Summarizing, the

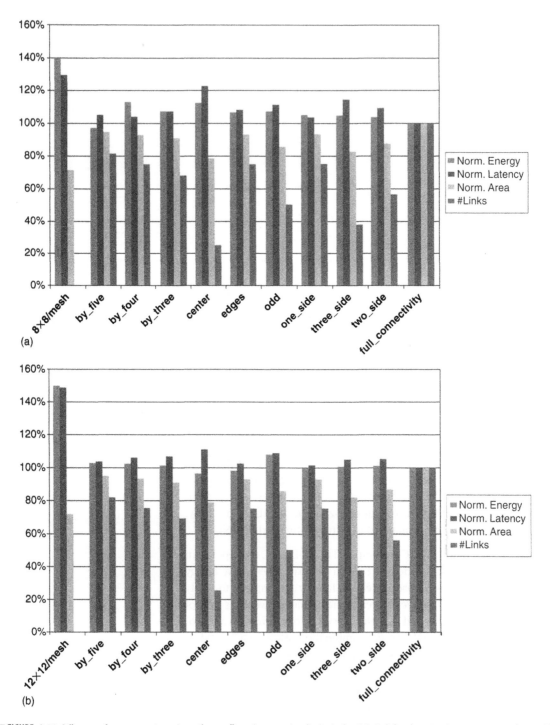

■ **FIGURE 9-24** Different performance metrics under uniform traffic and a normal traffic load of a 3-D NoC for alternative interconnection topologies with XYZ-OLD routing. (a) 64 network nodes and (b) 144 network nodes.

overall performance of a two-dimensional NoC is significantly lower, exhibiting an increased energy and latency of approximately 50%.

When the traffic load is increased by 50%, the performance of all of the router combinations degrades as compared to the fully connected 3-D NoC. This behavior occurs since in a 3-D NoC containing both 2-D and 3-D routers, a smaller number of 3-D routers is used to save energy, reducing, in turn, the interconnectivity within the network. This lower interconnectivity increases the number of hops required to propagate data packets, increasing the overall network latency.

In the case of low traffic loads, alternative combinations can be beneficial since the requirements for communication resources are low. The simulation results for a 64- and 144-node 2-D and 3-D NoC under low uniform traffic and **XYZ** routing are illustrated in Figure 9-25. An exception is the "edges" combination in the 64-node 3-D NoC (see Figure 9-25a), where all of the 3-D routers reside along the edges of each plane within the 3-D NoC. This arrangement produces a 7% increase in packet latency. The performance of the 2-D NoC again decreases with increasing NoC dimensions. This behavior is depicted in Figure 9-25b, where the 2-D NoC dissipates 38% more energy while the latency increases by 37%.

Finally, the energy and latency of the various interconnection patterns are compared to those of a fully connected 3-D NoC in Table 9-6. The three types of traffic are shown in the first column. The minimum and maximum value of the energy dissipation is listed in the next two columns. The minimum and maximum value of the average packet latency is reported in the fourth and fifth column, respectively. Although only the latency increases, which is expected as the alternative interconnection patterns decrease the interconnectivity of the network, certain traffic patterns produce a considerable savings in energy. This savings in energy is important due to the significance of thermal effects in 3-D circuits. In addition to on-chip networks, another design style that greatly benefits from vertical integration is the field programmable gate array (FPGA), which is the topic of the following section.

9.4 THREE-DIMENSIONAL FPGAs

Field programmable gate arrays are programmable integrated circuits that implement abstract logic functions with a considerably smaller design turnaround time as compared to other design styles, such as application-specific (ASIC) or full custom integrated circuits. Due to this flexibility, the share of the IC market for FPGAs has steadily

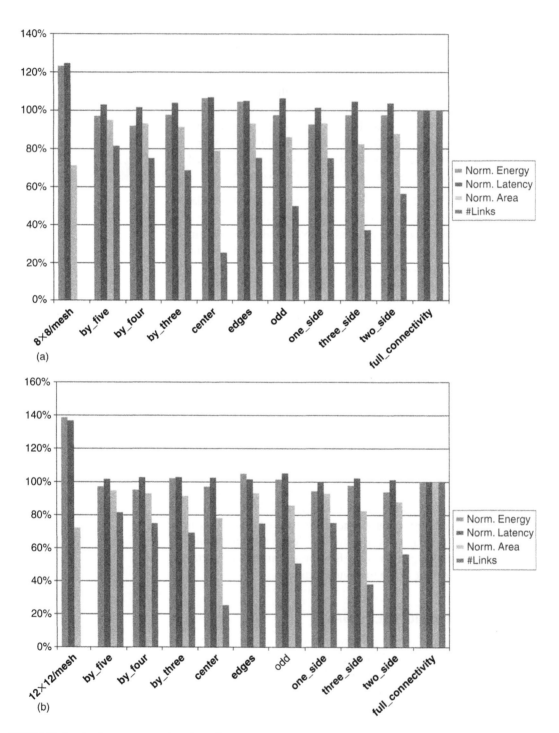

■ **FIGURE 9-25** Several performance metrics under uniform traffic and a low traffic load of a 3-D NoC for alternative interconnection topologies with **XYZ** routing: (a) a 4 × 4 × 4 3-D mesh and (b) a 6 × 6 × 4 3-D mesh.

Table 9-6 Min-max variation in 3-D NOC latency and power dissipation as compared to 2-D NOC under different traffic patterns and a normal traffic load.

| Traffic Pattern | Min-Max Energy and Latency Normalized | | | |
| | Energy | | Latency | |
	Min	Max	Min	Max
Uniform	92%	108%	98%	113%
Transpose	88%	116%	100%	354%
Hot spot	71%	116%	100%	134%

increased. The tradeoff for the reduced time to market and versatility of the FPGAs is lower speed and increased power consumption as compared to ASICs. A traditional physical structure of an FPGA is depicted in Figure 9-26, where the logic blocks (LBs) can implement any digital logic function with some sequential elements and arithmetic units [294]. The switch boxes (SBs) provide the interconnections among the logic blocks. The SBs include pass transistors, which connect (or disconnect) the incoming routing tracks with the outgoing routing

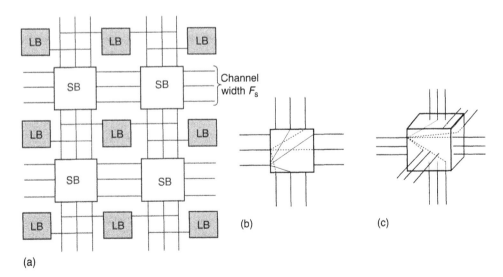

(a)

(b)   (c)

■ **FIGURE 9-26** Typical FPGA architecture: (a) a 2-D FPGA box, and (c) a 3-D switch box. A routing track can connect three outgoing tracks in a 2-D SB, while in a 3-D SB, a routing track can connect five outgoing routing tracks.

tracks. Memory circuits control these pass transistors and program the logic blocks for a specific application. In FPGAs, the SBs constitute the primary delay component of the interconnect delay between the logic blocks and can consume a great amount of power.

Extending FPGAs to the third dimension can improve performance while decreasing power consumption as compared to conventional planar FPGAs. A generalization of FPGAs to the third dimension would include multiple planar FPGAs, wafer or die bonded to form a 3-D system. The crucial difference between a 2-D and 3-D FPGA is that the SB provides communication to five logic blocks in a 3-D system rather than three neighboring logic blocks in a 2-D FPGA (see Figures 9-26b and 9-26c). Consequently, each incoming interconnect segment connects to five outgoing segments rather than three outgoing segments. The situation is somewhat different for the bottom and topmost plane of a 3-D FPGA, but in the following discussion this difference is neglected for simplicity. Since the connectivity of a 3-D SB is greater, additional pass transistors are required in each SB, increasing the power consumption, memory requirements to configure the SB, and, possibly, the interconnect delay. The decreased interconnect length and greater connectivity can compensate, however, for the added complexity and power of the 3-D SBs.

In order to estimate the size of the array beyond which the third dimension is beneficial, the shorter average interconnect length offered by the third dimension and the increased complexity of the SBs should be simultaneously considered [295]. Incorporating the hardware resources (*e.g.*, the number of transistors) required for each SB and the average interconnect length for a 2-D and 3-D FPGA, the minimum number of LBs for a 3-D FPGA to outperform a 2-D FPGA is determined from the solution of the following equation,

$$F_{s.2-D}\frac{2}{3}N^{1/2} = F_{s.3-D}N^{1/3}, \tag{9-27}$$

where $F_{s,2-D}$ and $F_{s,3-D}$ are the channel width of a 2-D and 3-D FPGA, respectively, and N is the number of logic blocks. Solving (9-27) yields $N = 244$, a number that is well exceeded in modern FPGAs.

Since the pass transistors, employed both in 2-D and 3-D SBs, contribute significantly to the interconnect delay, degrading the performance of an FPGA, those interconnects that span more than one LB can be utilized. These interconnect segments are named after the number of LBs that is traversed by these segments, as shown in Figure 9-27. Wires that span two, four, or even six LBs are quite common in

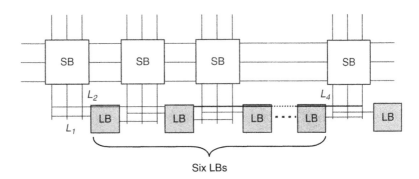

■ **FIGURE 9-27** Interconnects that span more than one logic block. L_i denotes the length of these interconnects, and i is the number of LBs traversed by these wires.

contemporary FPGAs. Interconnects that span one-fourth to a half of an IC edge are also possible [296].

The opportunities that the third dimension offers in SRAM-based 2-D FPGAs have also been investigated [297]. Analytic models that estimate the channel width in 2-D FPGAs have been extended to 3-D FPGAs. Hence, the channel width W for an FPGA with N LBs, exclusively consisting of unit-length interconnect segments and implemented in n physical planes, can be described by

$$W = \frac{\sum_{l=1}^{2\sqrt{N/n}-2+(n-1)d_v} lf_{3-D}(l)\chi_{\text{fpga}}}{\left(2N + \frac{(n-1)N}{n}\right)e_t}, \qquad (9\text{-}28)$$

where $f_{3-D}(l)$ is a stochastic interconnect length distribution similar to those discussed in Chapter 4. χ_{fpga} converts a point-to-point distance into an interconnect length, and e_t is the utilization parameter of the wiring tracks. These two factors can be determined from statistical data characterizing the placement and routing of benchmark circuits on FPGAs. Note that these factors depend on both the architecture of the FPGA and the automated layout algorithm used to route the FPGA.

Various characteristics of FPGAs have been estimated from benchmark circuits and randomized netlists placed and routed with the SEGment Allocator (SEGA) [298], the versatile place and route (VPR) [299] tools, and analytic expressions, such as (9-28). In these benchmark circuits, each FPGA is assumed to contain 20,000 four-input LBs and is implemented in a 0.25-μm CMOS technology. The area, channel density, and average wirelength are measured in LB pitches, which is the distance between two adjacent LBs, and are listed in Table 9-7 for different number of physical planes.

Table 9-7 Area, wirelength, and channel density improvement in 3-D FPGAs.

Number of Planes	Area [cm²]	Channel Density	Avg. Wirelength [LB Pitch]
1 (2-D)	7.84	41	8
2 (3-D)	3.1	24	6
3 (3-D)	1.77	20	5
4 (3-D)	1.21	18	5

■ **FIGURE 9-28** Interconnect delay for several number of physical planes: (a) average length wires and (b) die edge length interconnects.

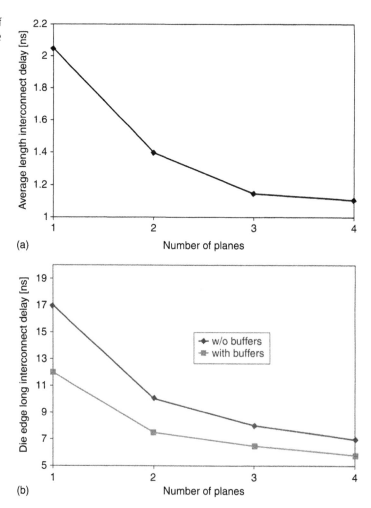

The improvement in the interconnect delay with a length equal to the die edge is depicted in Figure 9-28 for various number of planes. Those wires that span multiple LBs are implemented with unit length segments (*i.e.*, no SBs are interspersed along these wires), whereas the die edge long wires are implemented by interconnect segments with a length equal to a quarter of the die edge. A significant decrease in delay is projected; however, these gains diminish for more than four planes, as indicated by the saturated portion of the delay curves depicted in Figure 9-28. The components of the power dissipated in a 3-D FPGA assuming a 2.5-volt power supply are shown in Figure 9-29. The power consumed by the LBs remains constant since the structure of the logic blocks does not vary with the third dimension. However, due to the shorter interconnect length, the power dissipated by the interconnects is less. This improvement, however, is smaller than the improvement in the interconnect delay, as indicated by the slope of the curves illustrated in Figures 9-28a and 9-28b. This behavior is attributed to the extra pass transistors in a 3-D SB, which increase the power consumption, compromising the benefit of the shorter interconnect length. Due to the reduced interconnect length, the power dissipated by the clock distribution network is also less.

New design aids are required to exploit the latency and power benefits of FPGAs that stem from 3-D architectures. These CAD tools

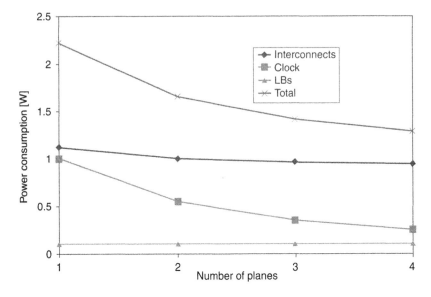

■ **FIGURE 9-29** Power dissipated by 2-D and 3-D FPGAs.

should have the ability to explore various architectures and interconnection schemes that can exist within a 3-D FPGA in addition to including appropriate physical design techniques. Several design aids that target different 3-D FPGAs are described in the following section.

9.4.1 Design Aids for 3-D FPGAs*

To date, several academic tools for the physical design of 3-D FPGAs have been developed, where the majority of these tools is based on existing tools developed for 2-D FPGAs. These 3-D oriented design aids typically include some exploratory capabilities and implement various physical design tasks, such as placement and routing.

A unified placement and routing system for 2-D FPGAs [300] has been extended to several 3-D FPGA architectures [301]. The system utilizes a top-down approach where the FPGA is successively partitioned into multiple cubic sections or cells until only one logic block is contained in each section [295]. This partition decomposes an FPGA into $n_x \times n_y \times n_z$ cells, where n_x and n_y are the number of cells in the x- and y-directions, respectively, and n_z is the number of planes within the 3-D FPGA. The LBs are arranged to minimize the number of nets that cross a cell boundary. A simulated annealing algorithm is employed to minimize the total interconnect length by swapping blocks among the cells. Upon completion of the placement process, a one-step routing algorithm that utilizes a Steiner-based heuristic generates the physical route of the circuit. This heuristic, which is based on [302], is generalized into three dimensions and iteratively applied to determine the individual route of the nets within a 3-D FPGA that minimizes the wirelength of the spanning tree.

This technique has been applied to a number of industrial benchmark circuits. The results are compared to a 2-D FPGA, which has been placed and routed with the Mondrian tool [300] followed by various postprocessing steps. The improvement in the average interconnect length, SB utilization, and average source-to-sink length of the paths (*i.e.*, radius) are 13.8%, 23.0%, and 26.4%, respectively, for a 3-D FPGA [301].

Another tool suitable for 3-D FPGAs is the three-dimensional place and route (TPR) program [303], which is also based on a tool

---

*This section, 9.4.1, was contributed by Professor Dimitrios Soudris of the Democritus University of Thrace.

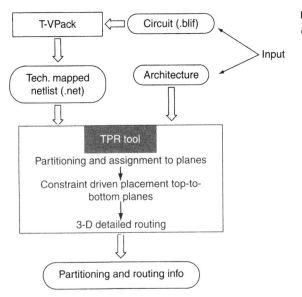

■ **FIGURE** 9-30 Design flow of a three-dimensional FPGA-based placement and routing tool [303].

developed for 2-D FPGAs, called VPR [299], [304]. An improved variant of the tool, called SA-TPR, includes a simulated annealing algorithm to explore the solution quality and computational time tradeoffs. A hybrid version that utilizes both the basic TPR and the SA-TPR has also been developed to generate more refined solutions.

Various steps of the TPR tool are illustrated in Figure 9-30. The input to the tool is a specific 3-D FPGA architecture and a circuit mapped onto this architecture described in .blif format. The placement commences by partitioning the blocks of the circuit into the planes of the stack to balance the resulting partitions. This initial partition is based on the `hmetis` algorithm [305], which minimizes the cut size among the planes or, equivalently, the number of interplane vias. This constraint is due to the low interplane via density or high fabrication cost of the vias. The assignment of the blocks within each plane follows, based on the approach described in [306]. In this approach, the blocks are assigned to ensure that the terminals of two connected blocks are aligned. In this case, a net with fewer interspersed SBs can be used to connect these blocks, resulting in a smaller interconnect delay. A L_4 segment can be used rather than four L_1 segments (see Figure 9-27). The routing step in TPR employs a rip-up and reroute technique that utilizes a breadth-first search algorithm [307].

Table 9-8 Improvement ratios normalized to the SA-TPR output for 2-D FPGAs.

Objective	SA-TPR 3-D Single Segment	SA-TPR 3-D Multi-segment	TPR 2-D	TPR 3-D Single Segment	TPR 3-D Multi-Segment	Hybrid 3-D Single Segment	Hybrid 3-D Multi-Segment
Delay	0.82	0.82	1.24	0.98	0.96	0.82	0.82
Wirelength	0.90	0.90	1.34	1.06	1.05	0.85	0.83
Routing area	1.11	1.05	1.24	1.34	1.30	1.18	1.13
Channel width (horizontal)	0.93	0.91	1.23	0.95	0.96	0.85	0.82

Adopting an SA engine for the placement algorithm, the placement solution decreases the total wirelength; however, the runtime of the algorithm is greatly increased. For the SA-TPR, the initial partitioning step is not necessary. This variant of the tool therefore explores a greater portion of the solution space as interplane moves among blocks placed on different planes are permitted at any iteration of the algorithm. In a hybrid version of the tool, a mixed partitioning scheme and simulated annealing algorithm are utilized to reduce the computational time without compromising the quality of the solution. In Table 9-8, different variations of the TPR tool are compared with the SA-TPR version applied to a 2-D FPGA considering various design objectives. The benefits that result from the third dimension are significant. The target architectures for these benchmark circuits are evaluated in terms of the gains obtained from employing different interconnect schemes in 3-D FPGAs.

The VPR and TPR tools, however, do not support a complete design flow starting from a hardware description language (HDL) description of an application to the generation of the binary bit stream to configure the FPGA. A recently developed toolset, called MEANDER, offers these capabilities and supports the use of different 3-D architectures, as illustrated in Figure 9-31[308].

As shown in Figure 9-31, specific steps of the 3-D MEANDER framework do not depend on the target FPGA architecture. Consequently, existing CAD tools for synthesis, technology mapping, and binary bit stream generation can be utilized for either 2-D or 3-D FPGAs. The physical design of 3-D FPGAs, however, requires substantially

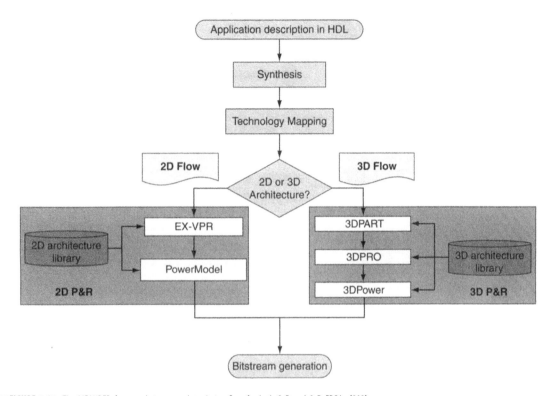

■ **FIGURE 9-31** The MEANDER framework is a complete design flow for both 2-D and 3-D FPGAs [309].

different approaches. New tools have therefore been developed, as depicted in the right-hand side of Figure 9-31 [308], [309].

These tools include exploratory capabilities where different partitions onto the planes of a 3-D FPGA and various SBs can be evaluated under different performance metrics, such as speed and power. The basic stages of the physical design process within the MEANDER framework are illustrated in Figure 9-32 and are as follows:

- Application partitioning
- Spatial allocation of the 3-D SBs for each plane of the 3-D FPGA
- Estimation of selected design metrics, such as delay and power
- Placement and routing of the application onto a target 3-D architecture.

Each of these stages consists of a number of tasks, as shown in Figure 9-32. As with other tools, during the initial stage the application is divided into a number of partitions at least equal to the number of physical planes supported by the target 3-D FPGA technology. The

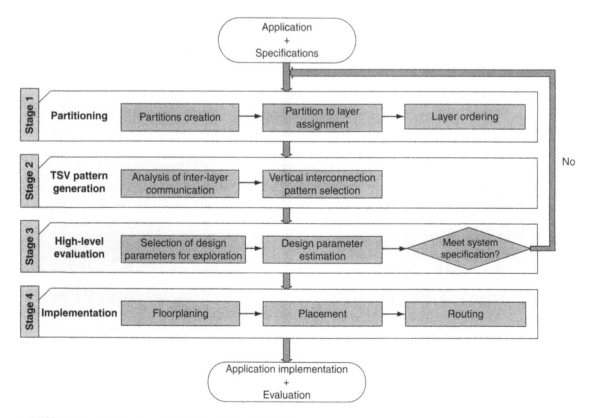

■**FIGURE 9-32** Physical design stages for 3-D FPGAs within the MEANDER framework.

primary criterion for choosing these partitions is to utilize most of the available logic and interconnect resources within a 3-D FPGA. The partitions are assigned to the planes of the 3-D FPGA, where more than one partition can be assigned to a physical plane. The assignment of the partitions to the planes is followed by ordering the planes. Different objectives can be used for this task. For example, two partitions, which include LBs that communicate frequently, can be placed on two adjacent physical planes to reduce the physical distance among these LBs. Alternatively, partitions with high switching activities should not be assigned to those planes located far from the heat sink.

A special feature of this tool is that different types and spatial distributions of the SBs can be investigated. The interconnect fabric is selected during the second stage, which includes two basic tasks. The required interplane connections among partitions on different planes are estimated to ensure that the necessary number of 3-D SBs is determined.

The spatial distribution of the 3-D SBs across each plane is determined in the second step.

The delay and power of the target 3-D FPGA architecture can be estimated based on the partitions and connectivity demands. These estimates are compared with the specified design objectives. If these goals are not satisfied, the design flow returns to the first stage to improve the partitions or change the target 3-D architecture and distribution of the 3-D SBs. If the design goals are satisfied, the final stage is initiated with the floorplanning, placement, and routing simultaneously progressing on all of the planes. Different cost functions, for example, timing or connectivity driven, can be adopted to guide these tasks.

A qualitative comparison of some of the tools discussed in this section is listed in Table 9-9. The 3DPRO tool exhibits superior characteristics as compared to TPR. These features include, for example, power-aware placement and routing. This objective can significantly lower thermal gradients in 3-D integrated systems. In addition, a large diversity of 3-D SBs is supported. Furthermore, different combinations of 2-D and 3-D SBs within a 3-D FPGA can be exploited to reduce the number of critical vertical interconnects. Furthermore, any type of switch box

Table 9-9 Qualitative comparison of CAD tools for 3-D FPGAs.

Feature	TPR [303]; [310]	3DPRO [308]
Delay estimation	Yes	Yes
Wirelength estimation	Yes	Yes
Power estimation	No	Yes
Supported switch boxes	Subset, Wilton, Universal	Designer specified
Heterogeneous interconnect (simultaneously 2D/3D SBs)	No	Yes
Architecture exploration	Yes	Yes
Timing-aware P&R	Yes	Yes
Power-aware P&R	No	Yes
Part of complete framework	No	Yes
Public available	Yes	Yes

can be employed, while the TPR tool is only compatible with the subset, Wilson, and universal types of switch boxes.

These design aids and related techniques assume a 3-D FPGA architecture in which each plane resembles a 2-D FPGA separately processed and stacked into a 3-D system. The primary difference is replacing the 2-D SBs with 3-D SBs that exploit the vertical channels. Another architectural approach assumes that the logic blocks and switch boxes are placed on separate physical planes [311]. A third plane that contains the memory required to configure the FPGA can also be used. With this 3-D FPGA architecture, the objective is to redistribute the basic components of an FPGA, namely, the logic blocks, switch boxes, routing resources, and configuration memory, to maximize the integration density within the 3-D stack. A reduction in area of approximately 70% of a conventional FPGA is achieved with a three-plane 3-D FPGA, where the memory and SBs are located on the upper two planes [311].

The VPR tool [299] places various circuits in a 2-D FPGA, and with simple RC delay and power models for the interconnect and pass transistors in the SBs, the delay and power consumption of conventional 2-D FPGAs can be estimated. These expressions have been extended to evaluate the interconnect length and delay of a 3-D FPGA. An appropriate scaling factor, smaller than one, for the interconnects within a 3-D FPGA is used to investigate the performance improvements originating from the three-plane architecture. For the 20 largest MCNC benchmark circuits [157], a possible three-plane FPGA architecture offers an improvement of 3.2, 1.7, and 1.7 times in logic density, critical path delay, and dynamic power consumption assuming a 65-nm CMOS technology [311].

9.5 SUMMARY

Different classes of circuit structures that can greatly benefit from three-dimensional integration are described in this chapter. Several architectures and related design aids for these circuits are discussed. The primary points of this discussion can be summarized as follows:

- Vertical integration can improve both the latency and power consumption of wire-limited and communication centric circuits. These circuits include the microprocessor-memory system, on-chip networks, and FPGAs.

- Implementing several components of a microprocessor across multiple physical planes decreases the power consumption and improves the speed by utilizing fewer redundant pipeline stages.

- Partitioning a cache memory into a 3-D structure can reduce the time required to access the memory.

- Stacking additional memory planes on a microprocessor can improve the overall performance of a microprocessor-memory system without exceeding the thermal budget of the system.

- 3-D NoCs are a natural evolution of 2-D NoCs and exhibit superior performance.

- The minimum latency and power consumption can be achieved in 3-D NoC by reducing both the number of hops per packet and the length of the communication channels.

- Expressions for the zero-load network latency and power consumption are described for 3-D NoCs.

- NoCs implemented by a 3-D mesh reduce the number of hops required to propagate a data packet between two nodes within a network.

- 2-D IC – 3-D NoC decrease the latency and power consumption by reducing the number of hops.

- 3-D IC – 2-D NoC decrease the latency and power consumption by reducing the buss length or, equivalently, the distance between adjacent network nodes.

- Large networks are primarily benefited by the 2-D IC – 3-D NoC topology, while small networks are enhanced by the 3-D IC – 2-D NoC topology.

- The distribution of nodes corresponding to either the minimum latency or power consumption of a network depends on both the interconnect parameters and network size.

- A tradeoff exists between the number of planes utilized to implement a network and the number of PEs. Consequently, and not surprisingly, the 3-D IC – 3-D NoC topology achieves the greatest improvement in latency and power consumption by most effectively exploiting the third dimension.

- A simulator that supports the exploration of 3-D NoC topologies with different routing schemes, traffic patterns, and traffic loads has been developed.

- A multiplane on-chip network that includes both 2-D and 3-D routers can produce a significant savings in energy and area with a tolerable increase in latency as compared to a full 3-D topology.

- Three-dimensional FPGAs have been proposed to improve the performance of 2-D FPGAs, where multiple planar FPGAs are bonded to form a 3-D stack of FPGAs.

- A critical issue in 3-D FPGAs is the greater complexity of the 3-D switch box that can negate the benefits from the shorter interconnect length and enhanced connectivity among the logic blocks.

Case Study: Clock Distribution Networks for 3-D ICs

In the previous chapters, the advantages that result from introducing the third dimension to integrated circuit applications are discussed. Significant improvement in interconnect performance and power consumption is predicted for these circuits, constituting an important opportunity for high-performance digital circuits. An omnipresent and challenging issue for synchronous digital circuits is the reliable distribution of the clock signal to the many hundreds of thousands of sequential elements distributed throughout a synchronous circuit [312]. The complexity is further increased in 3-D ICs as sequential elements belonging to the same clock domain (*i.e.*, synchronized by the same clock signal) can be located on different planes. Another important issue in the design of the clock distribution network is low-power consumption, as the clock network dissipates a significant portion of the total power consumed by a synchronous circuit [313], [314]. This demand is stricter for 3-D ICs due to the increased power density and related thermal limitations, as discussed in Chapter 6.

In 2-D circuits, symmetric interconnect structures, such as H-trees and X-trees, are widely utilized to distribute the clock signal across a circuit [315]. The symmetry of these structures permits the clock signal to simultaneously arrive at the leaves of the tree, resulting in synchronous data processing. Maintaining this symmetry within a 3-D circuit, however, is a difficult task. In addition, 3-D ICs include two different types of interconnects, namely, intraplane and interplane interconnects that can exhibit different impedance characteristics [316]. Furthermore, interplane interconnects have fixed dimensions determined by the target 3-D technology, further constraining the design space.

In this chapter, a variety of clock network architectures for 3-D circuits are investigated. These clock topologies have been included on a test circuit and manufactured in the 3-D technology developed at MIT Lincoln Laboratories (MITLL). This fabrication process is discussed in the following section. The logic circuitry comprising the common load of the 3-D clock distribution networks is described in Section 10.2. The various clock distribution networks that have been employed in this case study are described in Section 10.3. Experimental results and a comparison of the different clock distribution networks are presented in Section 10.4. A short summary is provided in the last section of the chapter.

10.1 MIT LINCOLN LABORATORIES 3-D IC FABRICATION TECHNOLOGY

The MIT Lincoln Laboratories recently developed a manufacturing process for fully depleted silicon-on-insulator (FDSOI) 3-D circuits with short interplane vias (also called 3-D vias here for simplicity). The most attractive feature of this process is the high density of the 3-D vias as compared to other 3-D technologies currently under development, as reviewed in Chapter 3. The MITLL process is a wafer-level 3-D integration technology with up to three FDSOI wafers bonded to form a 3-D circuit. The diameter of the wafers is 150 mm. The minimum feature size of the devices is 180 nm, with one polysilicon layer and three metal layers interconnecting the devices on each wafer. A backside metal layer also exists on the upper two planes, providing the starting and landing pads for the 3-D vias, and the I/O, power supply, and ground pads for the entire 3-D circuit. The primary steps of this fabrication process are illustrated in Figures 10-1 to 10-6 [140], [221].

Each of the wafers is manufactured by a mainstream FDSOI process (Figure 10-1). The second wafer is flipped and front-to-front bonded, with the first wafer using oxide bonding (Figure 10-2). The handle wafer is removed from the second wafer, and the 3-D vias are etched through the oxide of both planes. Tungsten is deposited to fill the 3-D vias, and the surface of the wafer is planarized by chemical mechanical polishing (Figure 10-3). The backside vias and metallization are formed to provide the pads for the 3-D vias of the third wafer and these vias are interconnected with the M1 layer of the second wafer (Figure 10-4). The third wafer is also flipped and front-to-back bonded with the second plane (Figure 10-5). Another etching step is used to form the 3-D vias of the third plane. The backside vias and interconnections are formed along with the I/O and power pads. A glass layer provides the passivation layer, while the overglass cuts create the necessary pad openings for the off-chip interconnections (Figure 10-6).

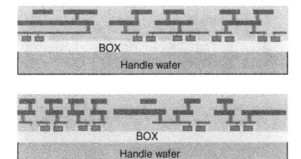

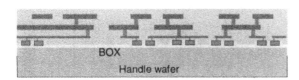

■ **FIGURE 10-1** Three wafers are individually fabricated with an FDSOI process.

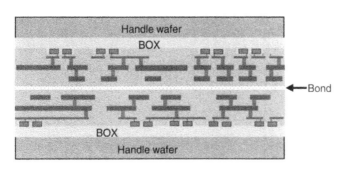

■ **FIGURE 10-2** The second wafer is front-to-front bonded with the first wafer.

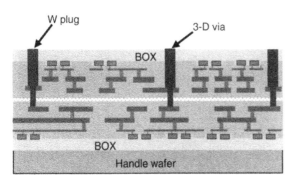

■ **FIGURE 10-3** The 3-D vias are formed, and the surface is planarized with CMP.

■**FIGURE 10-4** The backside vias are etched, and the backside metal is deposited on the second wafer.

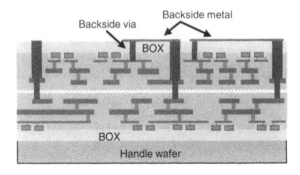

■**FIGURE 10-5** The third wafer is front-to-back bonded with the second wafer, and the 3-D vias for that plane are formed.

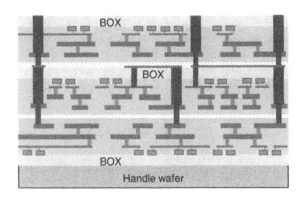

■**FIGURE 10-6** Backside metal is deposited, and glass layer is cut to create openings for the pads.

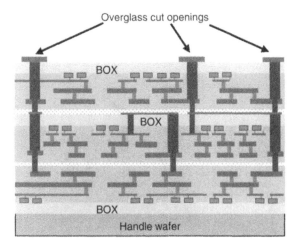

A salient characteristic of this process is the short 3-D vias. As illustrated in Figure 10-7, the total length of a 3-D via that connects two devices on the first and third planes is approximately 20 µm. In addition, the dimensions of these vias are 1.75 µm × 1.75 µm, much smaller than the size of the through silicon via in many existing 3-D technologies, as discussed in Chapter 3. The spacing among the 3-D vias depends on the density of these vias and ranges from 1.75 µm to 8 µm. Note that the 3-D vias connecting the second and third planes can be vertically stacked, resulting in 3-D vias that directly connect devices on the first and third planes.

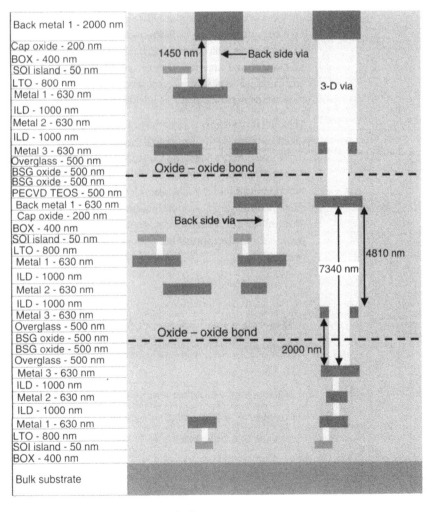

■**FIGURE 10-7** Layer thicknesses in the 3-D IC MITLL technology [140].

A doughnut-shaped structure is required for the 3-D vias in both the second and third planes to provide mechanical support for these vias. As shown in Figure 10-7, the 3-D vias connect different metal layers in the second and third planes. A 3-D via for the second plane connects the backside metal with the M3 layer of the first plane through the doughnut formed by M3 in the second plane. The backside vias or the M3 layer of the second plane can be utilized to connect the 3-D via with the devices on that plane. Alternatively, a 3-D via for the third plane starts from the backside metal of the third plane and ends on the backside metal of the second plane through a doughnut formed by the M3 layer of the third plane. The backside vias connect this via with the devices on the second plane. The transistors located on the third plane are connected to the 3-D via either through the backside metal layer and backside vias, or through the M3 doughnut that surrounds the 3-D via. Note that the 3-D vias can be placed anywhere within the circuit and not only within certain regions. The minimum distance from the transistors, however, is specified by the design rules.

The electrical sheet resistance of the metal and diffusion layers is listed in Table 10-1 along with the bulk resistivity. The total resistance of the intraplane vias and contacts is listed in Table 10-2. Since the third plane is also intended for RF circuits, a low-resistance back-side metal is available. In addition to the active devices, passive elements, such as resistors and metal-insulator-metal (MIM) capacitors, are also available. The resistors, however, can only be placed on the third plane where the polysilicon or active layer is utilized to form these resistors.

Table 10-1 Layer resistances of the 3-D FDSOI process [140].

Parameter	Value
Bulk resistivity	$\sim$2000 Ω-cm
Silicided n^+/p^+ active sheet resistance	15 $\pm$ 3 Ω/sq
Silicided n^+/p^+ polysilicon sheet resistance	15 $\pm$ 3 Ω/sq
Silicided n^+/p^+ sheet resistance	15 $\pm$ 3 Ω/sq
Lower metal layer sheet resistance	$\sim$0.12 Ω/sq
Top metal layer sheet resistance	$\sim$0.08 Ω/sq
Backside metal sheet resistance	$\sim$0.12 Ω/sq

Table 10-2 Contact and via resistances of the 3-D FDSOI process [140].

Parameter	Value
Poly contact (250 nm × 250 nm)	10 ± 2 Ω
n^+ active contact (250 nm × 250 nm)	10 ± 2 Ω
p^+ active contact (250 nm × 250 nm)	10 ± 2 Ω
Interconnect metal via (300 nm × 300 nm)	4 Ω
Backside metal via (500 nm × 500 nm)	2 Ω

In support of this manufacturing process, design kits for this technology have been developed by academic institutions and supporting computer aided design CAD companies. The process design kit has been developed for the Cadence Design Framework by faculty from North Carolina State University [188]. This kit includes device models for circuit simulation and a sophisticated layout tool for 3-D circuits. For example, circuits located on a specific plane can be separately visualized or highlighted. In addition, a circuit designed for one plane can be reproduced or transferred to another plane. Complete design rule checking and circuit extraction are also available. Electrical rule checking, however, is not included, meaning that these 3-D circuits cannot be directly checked for shorts between the power and ground lines. However, to mitigate this problem, two pins can be assigned different names on the power and ground lines. Design rule checking reports errors whenever lines with different names are crossed, thereby checking for electrical shorts.

10.2 **3-D CIRCUIT ARCHITECTURE**

A test circuit exploring a variety of clock network topologies suitable for 3-D ICs has been designed based on the process described in the previous section. A block diagram of the circuit is depicted in Figure 10-8. The test circuit consists of four blocks. Each block contains the same logic circuit but implements a different clock distribution network. The total area of the test circuit is 3 mm × 3 mm. The logic circuit common to all of these blocks is described in this section while the various clock network topologies are discussed in Section 10.3.

An overview of the logic circuitry is depicted in Figure 10-9. The function of this logic is to emulate different switching patterns characterizing

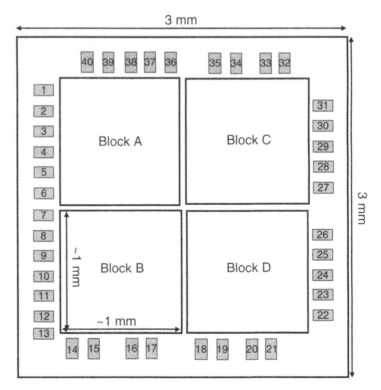

■ **FIGURE 10-8** Block diagram of the 3-D test IC. Each block has an area of approximately 1 mm². The remaining area is reserved for the I/O pads (the grey rectangles).

the circuit and load conditions for the clock distribution networks under investigation. The logic is repeated in each plane and includes

- Pseudorandom number generators
- A crossbar switch
- Control logic for the crossbar switch
- Groups of 4-bit counters
- Current loads and an output circuit for probing

The pseudorandom number generators are based on the technique described in [317], which uses linear feedback shift registers and XOR operations to generate a random 16-bit word every clock cycle after the first few cycles required to initialize the generator. The physical layout of one random number generator is illustrated in Figure 10-10. A total of nine pseudorandom generators are used in each circuit block, connected by groups of three to the crossbar switch within each plane.

A classic crossbar switch with six input and output ports is included in each plane, where the width of each port is 16 bits. Three of the six

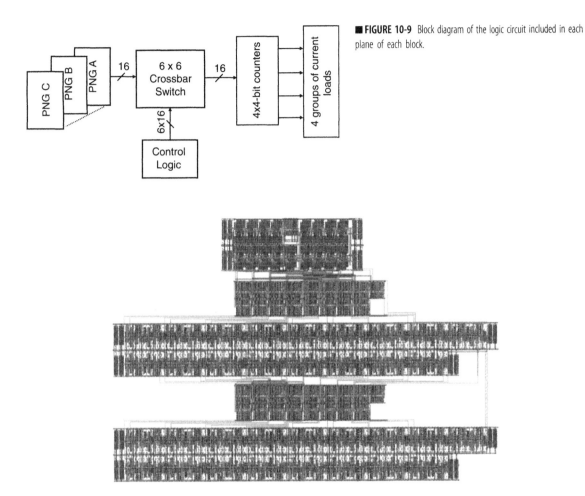

■ **FIGURE 10-9** Block diagram of the logic circuit included in each plane of each block.

■ **FIGURE 10-10** Physical layout of the pseudorandom number generator.

inputs of the crossbar switch are connected to the output of the number generators, while the remaining inputs are connected to ground. The physical layout of the crossbar switch is shown in Figure 10-11. The three output ports of the switch are connected to a group of four-bit counters, while the remaining outputs drive a small capacitive load. Since each port is 16 bits wide, each port is connected to four 4-bit counters. These counters are, in turn, connected to current loads implemented with cascoded current mirrors. The counters and current loads are distributed across each plane. The control logic consists of an 8-bit counter that controls the connectivity among the input and output ports of the crossbar switch.

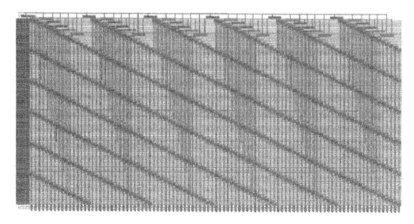

■ **FIGURE 10-11** Physical layout of the 6 × 6 crossbar switch with 16-bit wide ports.

The data flow in this circuit can be described as follows. After resetting the circuit, the pseudorandom number generators are initialized, and the control logic connects each input port to the appropriate output port. Since the control logic includes an 8-bit counter, each input port of the crossbar switch is successively connected every 256 clock cycles to each output port.

The output ports of the crossbar switch are connected to the 4-bit counters. Each of these counters is loaded with a 4-bit word, counts upward, and loaded with a new word every time all of the bits are equal to one. The MSB of each counter is connected to four current loads that are turned on when the bit is equal to one. Since the counters are loaded with random words from the random generators through the crossbar switch, the current loads draw a variable amount of current during circuit operation. This randomness is used to mimic different switching patterns that can exist within a circuit.

The current loads are implemented with cascoded current mirrors, as shown in Figure 10-12. In the cascoded current mirrors, the output current I_{out} closely follows I_{ref} as compared to a simple current mirror. The reference current I_{ref} is externally provided to control the amount of current drawn from the circuit. The gate of transistor M5 is connected to the MSB of a 4-bit counter, shown in Figure 10-12 as the *sel* signal. This additional device is used to switch the current sinks. The layout of a group of four current loads is illustrated in Figure 10-13. The width of the devices shown in Figure 10-12 is

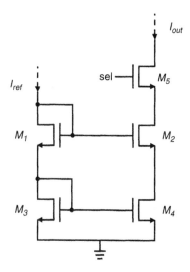

■ **FIGURE 10-12** Cascoded current mirror with an additional control transistor.

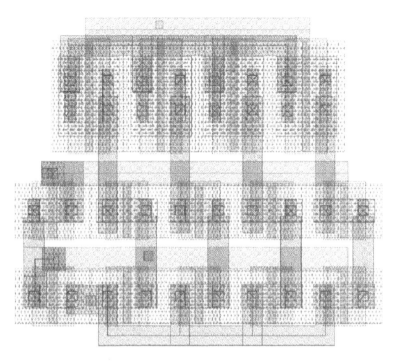

$$W_1 = W_2 = W_3 = W_4 = 600\,\text{nm}, W_5 = 2000\,\text{nm}. \qquad (10\text{-}1)$$

The power supply is 1.5 volts, as set by the MITLL process.

The layout of the test circuit is illustrated in Figure 10-14, while the connectivity of the pads to the external signals is listed in Table 10-3. Several decoupling capacitors are included in each circuit block and are highlighted in Figure 10-14. The capacitors serve as extrinsic decoupling capacitance and are implemented by MIM capacitors [175]. Note that the number of pads is not limited by the area of the circuit but rather by the maximum number of connections permitted by the available probe card. The backside metal layer of the third plane is utilized for the pads. Each of the circuit blocks is supplied by separate power and ground pads, while only a pair of power and ground pads is connected to the pad ring to provide protection from electrostatic discharge.

10.3 **CLOCK SIGNAL DISTRIBUTION IN 3-D CIRCUITS**

Different clock distribution network architectures for 3-D circuits are discussed in this section. Fundamental concepts for the timing of synchronous systems are introduced. Some design issues related to

Table 10-3 Pad connectivity of the 3-D test circuit (pad index shown in Figure 10-8).

Index	Pad Connectivity	Index	Pad Connectivity
1	reset	21	reset
2	V_{dd}	22	V_{ss}
3	V_{ss}	23	V_{dd}
4	Output bit	24	Output bit
5	V_{ss}	25	V_{dd}
6	V_{dd}	26	V_{ss}
7	Output bit	27	V_{ss}
8	reset	28	V_{dd}
9	V_{ss}	29	I_{ref}
10	V_{dd}	30	V_{ss}
11	I_{ref}	31	V_{dd}
12	V_{ss}	32	reset
13	V_{dd}	33	V_{dd}
14	V_{dd}	34	V_{ss}
15	V_{ss}	35	Output bit
16	V_{dd}	36	V_{ss}
17	V_{ss}	37	V_{dd}
18	I_{ref}	38	I_{ref}
19	V_{dd}	39	V_{ss}
20	V_{ss}	40	V_{dd}

distributing a clock signal in synchronous circuits are also discussed in Section 10.3.1. The clock networks employed in each of the blocks within the test circuit are described in Section 10.3.2.

10.3.1 Timing Characteristics of Synchronous Circuits

In synchronous circuits, the clock signal provides a common time reference for all of the sequential elements, orchestrating the flow of the data signals within a circuit [312]. A number of clock network

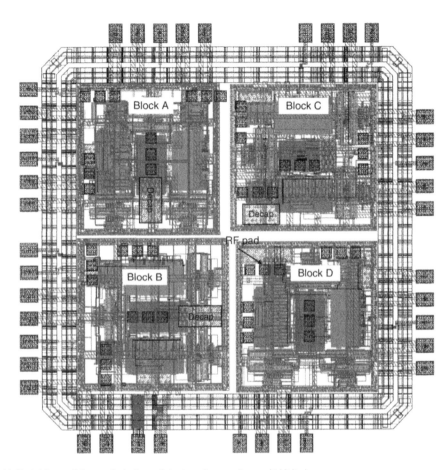

■ **FIGURE 10-14** Physical layout of the test circuit. Some of the decoupling capacitors are highlighted.

topologies have been developed for 2-D circuits, which can be symmetric, such as H-trees and X-trees, highly asymmetric, such as buffered tress and serpentine shaped structures [318], [319], and gridlike structures, such as rings and meshes. The clock distribution network is structured as a global network with multiple smaller local networks. Within the global clock network, the clock signal is distributed to specific locations across the circuit. These locations are the source of the local networks that pass the clock signal to the registers or other storage elements.

A symmetric structure such as an H-tree is often utilized in global clock networks [319], as shown in Figure 10-15. The most attractive

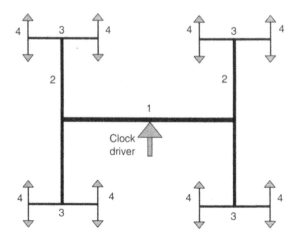

■**FIGURE 10-15** Two-dimensional four level H-tree.

■**FIGURE 10-15** Two-dimensional four level H-tree.

characteristic of symmetric structures is that the clock signal ideally arrives simultaneously at each leaf of the clock tree. Due to several reasons, however, such as load imbalances, process variations, and crosstalk, the arrival time of the clock signal at various locations within a symmetric tree can be different, producing clock skew. More precisely, clock skew is defined as the difference between the clock signal arrival time of *sequentially adjacent* registers [312]. Two registers are *sequentially adjacent* if these registers are connected with combinatorial logic (*i.e.,* no additional registers intervene between *sequentially adjacent* registers). An illustration of this data path is provided in Figure 10-16. An expression for the clock skew T_{skew} between these registers is also shown in Figure 10-16, where T_{C_i} and T_{Cj} are the arrival times of the clock signal at register R_i and R_j, respectively.

The clock signal typically traverses long distances to reach each sequential element within a circuit. In symmetric structures, such as an H-tree, the traversed distances are often longer to preserve symmetry. Due to these extremely long interconnects and high clock frequencies, the clock network is typically modeled as a transmission line, since inductive behavior is likely to occur [278]. Such behavior can cause multiple reflections at the branch points, directly affecting the speed and power consumed by the clock network. In order to lessen the reflections at the branch points of the tree, the interconnect width of the segments at each branch point is halved (for a 2× change in the line width) to ensure that the total impedance seen at that branch point is maintained constant (matched impedance).

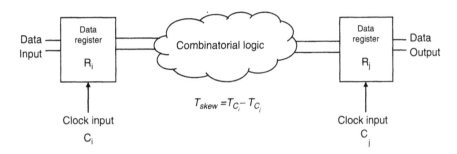

■ **FIGURE 10-16** A data path depicting a pair of *sequentially adjacent* registers.

10.3.2 **Clock Distribution Network Structures within the Test Circuit**

Different clock distribution schemes for 3-D circuits are described in this subsection. To evaluate the specific requirements of the clock networks, consider a traditional H-tree topology. As shown in Figure 10-15, at each branch point of an H-tree, two branches emanate with the same length. An extension of the H-tree to three dimensions does not guarantee equidistant interconnect paths from the source to the leaves of the tree. This situation is shown in Figure 10-17, where an H-tree is replicated in each plane of a 3-D circuit. The clock signal is propagated through interplane vias from the output of the clock driver to the center of the H-tree on planes one and three. The impedance of these vias increases the time for the clock signal to arrive at the leaves of the tree on these planes as compared to the time for the clock signal to arrive at the leaves of the tree on the same plane as the clock driver. Furthermore, in a multiplane 3-D circuit, three or four branches can emanate

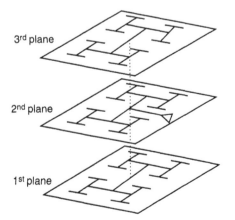

3rd plane

2nd plane

1st plane

■ **FIGURE 10-17** Two-dimensional H-trees constituting a clock distribution network for a 3-D IC.

from each branch point. The third branch propagates the clock signal to the other planes of the 3-D circuit, as shown in Figure 10-17. Similar to a design methodology for a 2-D H-tree topology, the width of each branch is reduced to a third (or more) of the segment preceding the branch point in order to match the impedance at the branch point. This requirement, however, is difficult to achieve as the third and fourth branches are connected by an interplane via. The vertical interconnects are of significantly different length as compared to the horizontal branches and also exhibit different impedance characteristics.

Various clock network topologies for 3-D ICs are investigated in this case study. Each of the four blocks of the test circuit includes a different clock distribution structure, and all are schematically illustrated in Figure 10-18. The physical layout of these topologies for the MITLL 3-D technology is depicted in Figure 10-19. The architectures employed in the blocks are as follows:

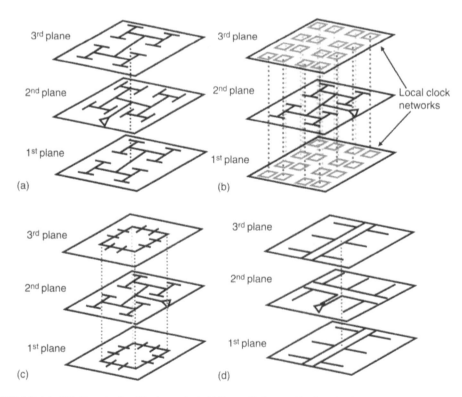

■**FIGURE 10-18** 3-D clock distribution networks within the test circuit: (a) H-trees, (b) H-tree and local rings/meshes, (c) H-tree and global rings, and (d) trunk based.

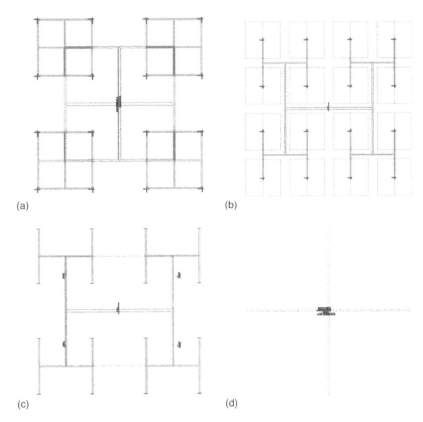

■ **FIGURE 10-19** Physical layout of the 3-D clock distribution networks: (a) H-trees, (b) H-tree and local rings/meshes, (c) H-tree and global rings, and (d) trunk based.

Block A: All of the planes contain a four level H-tree (*i.e.*, equivalent to 16 leaves) with identical interconnect characteristics. The H-trees are connected through a group of interplane vias, as shown in Figure 10-18a. The second plane is front-to-front bonded with the first plane, and both of the H-trees are implemented on the third metal (M3) layer. The physical distance between these clock networks is approximately 2 μm. Note that the H-tree on the second plane is rotated by 90° with respect to the H-trees on the other two planes. The orthogonal placement of these two clock networks effectively eliminates the inductive coupling. All of the H-trees are shielded with two parallel lines connected to ground.

Block B: A four-level H-tree is included in the second plane. Each of the leaves of this H-tree is connected through interplane vias to small local rings on the first and second plane, as illustrated in Figure 10-18b. As in Block A, the

H-tree is shielded with two parallel lines connected to ground. Additional interconnect resources are used to form local meshes. Due to the limited interconnect resources, however, a uniform mesh in each ring is difficult to achieve. The clock routing is constrained by the power and ground lines as only three metal layers are available on each plane.

Block C: The clock distribution network for the second plane is a shielded four-level H-tree. Two global rings are utilized for the other two planes, as depicted in Figure 10-18c. Each ring is connected through interplane vias to the four branch points on the second level of the H-tree. The registers in each plane are individually connected to the ring.

Block D: The clock network on each plane consists of a trunk structure and branches that connect the registers in each plane to the trunk, as shown in Figure 10-18d. As for Block A, the trunk for the second plane is rotated by 90° to avoid inductive coupling. Those interconnects that branch from the trunk are placed as close as possible to the registers.

Buffers are inserted at appropriate branch points within the H-trees to amplify the clock signal. In each of the circuit blocks, the clock driver for the entire clock network is located on the second plane. The clock driver on that plane is placed to ensure that the clock signal propagates through similar vertical interconnect paths to the first and third planes, resulting in the same effective delay for the registers located on the first and third planes. The clock driver is implemented with a traditional chain of tapered buffers [320]–[323].

The clock network on each plane feeds the registers located on the same plane. The off-chip clock signal is passed to the clock driver through an RF pad, as shown in Figure 10-20. The dimensions of the RF pad are also shown in Figure 10-20. Additional RF pads are placed at different locations on the third plane of each block for probing. These RF pads are used to measure the clock skew at different locations on different planes within the clock network. The output circuitry is an open drain transistor connected to the RF pads by a group of interplane vias to decrease the resistance between the transistor and the probe. The probe is modeled as a series RLC impedance with the values shown in Figure 10-21. The circuit depicted in Figure 10-21 is used to determine the size of the output transistor.

In addition to the clock skew of the clock network topologies employed within the blocks of the test circuit, the power consumption

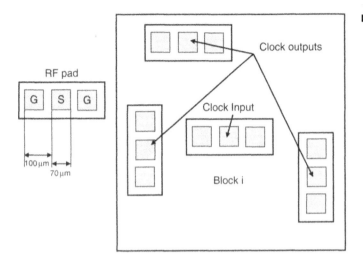

■ **FIGURE 10-20** Clock signal probes with RF pads.

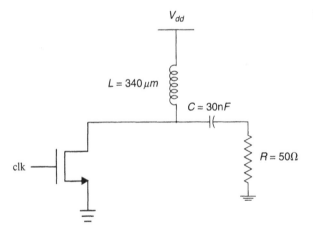

■ **FIGURE 10-21** Open drain transistor and circuit model of the probe.

of the entire clock distribution network has also been measured. Since all of the blocks include the same logic circuits, any difference in power consumption is attributed to the clock network, including the interconnect structures, clock driver, and clock buffers. The measurements of the clock skew and power dissipation of the blocks are discussed in the following section.

10.4 **EXPERIMENTAL RESULTS**

The experimental results of the clock distribution network topologies of the 3-D test circuit are evaluated in this section. The fabricated circuit is

■ **FIGURE 10-22** Top view of the fabricated 3-D test circuit.

■ **FIGURE 10-22** Top view of the fabricated 3-D test circuit.

depicted in Figure 10-22, where the four individual blocks can be distinguished. A magnified view of one block is shown in Figure 10-23. Each block includes four RF pads for measuring the delay of the clock signal. The pad located at the center of each block provides the input clock signal. The clock input is a sinusoidal signal with a DC offset, which is converted to a square waveform at the output of the clock driver. The remaining three RF pads are used to measure the delay of the clock signal at specific points on the clock distribution network within each plane. A buffer is connected to each of these measurement points. The output of this buffer drives the gate of an open drain transistor connected to the RF pad. The RF probes landing on these pads are depicted in Figure 10-24, where the die assembly on the probe station is illustrated.

A clock waveform acquired from the topology combining an H-tree and global rings, shown in Figure 10-18c, is illustrated in Figure 10-25, demonstrating operation of the circuit at 1.4 GHz. The clock skew between the planes of each block is listed in Table 10-4. The topologies are ordered in Figure 10-26 in terms of the maximum measured clock skew between two planes. The delay of the clock signal from the RF input pad at the center of each block to the measurement point on plane

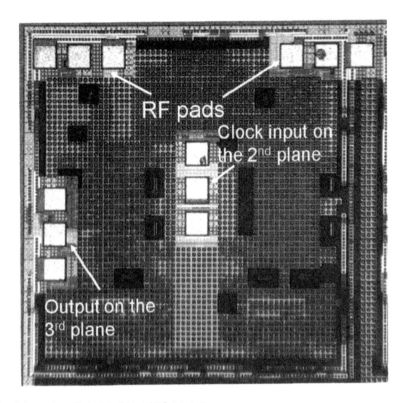

■ **FIGURE 10-23** Magnified view of one block of the fabricated 3-D test circuit.

■ **FIGURE 10-24** Die assembly of the 3-D test circuit with RF probes.

■ **FIGURE 10-25** Clock signal input and output waveform from the topology illustrated in Figure10-18c.

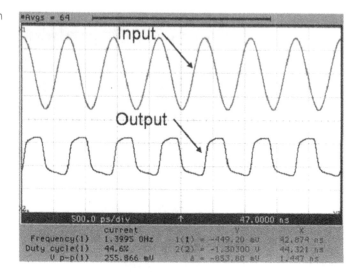

Table 10-4 Measured clock skew among the planes of each block.

Clock Distribution Network	Clock Skew [ps]		
	$Tc_{(B\text{-}A)} = Tc_B - Tc_A$	$Tc_{(B\text{-}C)} = Tc_{(B} - Tc_{C)}$	$Tc_{(A\text{-}C)} = Tc_A - Tc_C$
H-trees (Figure10-18a)	32.5	28.3	-4.2
Local meshes (Figure10-18b)	-68.4	-18.5	49.8
Global rings (Figure10-18c)	-112.0	-130.6	-18.6

i is denoted as Tc_i in Table 10-4. For example, Tc_A denotes the delay of the clock signal to the measurement point on plane A. In addition, the difference in the delay of the clock signal between two measurements points on planes i and j is notated as $Tc_{i\text{-}j}$.

For the H-tree topology, the clock signal delay is measured from the root to a leaf of the tree on each plane, with no other load connected to these leaves. The skew between the leaves of the H-tree on planes A and C (*i.e.*, $Tc_{A\text{-}C}$) is effectively the delay of a stacked 3-D via traversing the three planes to transfer the clock signal from the target leaf to the RF pad on the third plane. The delay of the clock signal to the sink of the H-tree on the second plane Tc_B is larger due to the additional capacitance within that quadrant of the H-tree. This capacitance is intentional on-chip decoupling capacitance placed under the quadrant, increasing the

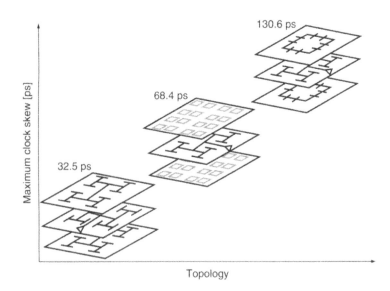

measured skew of Tc_{B-C} and Tc_{B-A}. This topology produces, on average, the lowest skew as compared to the two other topologies.

In the H-tree topology, each leaf of a tree is connected to registers located only within the same plane. Allowing one sink of an H-tree to drive a register on another plane adds the delay of another 3-D via to the clock signal path, further increasing the delay. Consequently, the registers within each plane are connected to the H-tree on the same plane. Note that this approach does not imply that these registers only belong to data paths contained within the same plane.

The clock skew among the planes is greater for the local mesh topology as compared to the H-tree topology, primarily due to the imbalance in the clock load for certain local meshes. Indeed, this topology has only 16 tap points within the global clock distribution network; three times fewer than the H-tree topology illustrated in Figure 10-18a. This difference can produce a considerable load imbalance, greatly increasing the local clock skew as compared to the local clock skew within the H-tree topology. By inserting the local meshes on planes A and C, which are connected to the 16 sinks of the H-tree on the second plane, the local clock skew is smaller. The greatest difference in the load is between the measurement points on planes A and B, which also produces the largest skew for this topology. The increase in skew, however, as compared to the H-tree topology, is moderate.

Consequently, a limitation of the local meshes topology is that greater effort is required to control the local skew. The lower number of sinks driven by the global clock distribution network increases the number of registers clocked by each sink. To better explain this situation, consider a segment of each topology shown in Figures 10-27a and 10-27b, respectively. For the H-tree topology, the clock signal is distributed from three sinks, one on each plane, to the registers within the circular area depicted in Figure 10-27a. Note that the radius of the circle on planes A and C is slightly smaller to compensate for the additional delay of the clock signal caused by the impedance characteristics of the 3-D vias. The registers located within these regions satisfy specific local skew constraints. Alternatively, in the case of the local mesh topology, the clock signal at the sinks of the H-tree on the second plane feeds registers in each of the three planes. Consequently, each sink of the tree connects to a larger number of registers as compared to the H-tree topology, as depicted by the shaded region in Figure 10-27b. Despite the beneficial effect of the local meshes, load imbalances are more pronounced for this topology. Alternatively, the H-tree topology (see Figure 10-18a) utilizes a significant amount of interconnect resources, dissipating more power.

The clock distribution network with the global rings exhibits low skew for planes A and C, those planes that include the global rings. The objective of this topology is to evaluate the effectiveness of a less symmetric architecture in distributing the clock signal within a 3-D circuit. Although the clock load on each ring is nonuniformly distributed, the load balancing characteristic of the rings yields a relatively low skew between the planes. Since the clock distribution network on the second plane is implemented with an H-tree, the skew

■ **FIGURE 10-27** Part of the clock distribution networks illustrated in Figures 10-18a and 10-18b. (a) The local clock skew is individually adjusted within each plane for the H-tree topology, and (b) the local skew is simultaneously adjusted for all of the planes for the local mesh topology.

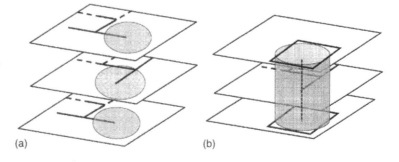

(a)   (b)

between adjacent planes is significantly larger than the skew between the top and bottom planes. Note that the sinks of the H-tree are located at a great distance from the rings on planes A and C (see Figure 10-18c). Consequently, a combination of H-tree and global rings is not a suitable approach for 3-D circuits due to the difficulty in matching the distance that the clock signal traverses on each plane from the sink of the tree or the ring to the many registers distributed across a plane.

The measured power consumption of the blocks operating at 1 GHz is reported in Table 10-5. An ordering of the blocks in terms of the measured dissipated power is illustrated in Figure 10-28. The local mesh topology dissipates the lowest power. This topology requires the least interconnect resources for the global clock network, since the local meshes are connected at the output of the buffers located

Table 10-5 Measured power consumption of each block operating at 1 GHz.

Clock Distribution Network	Power Consumption [mW]
H-trees (Figure10-18a)	260.3
Local meshes (Figure10-18b)	168.3
Global rings (Figure10-18c)	228.5

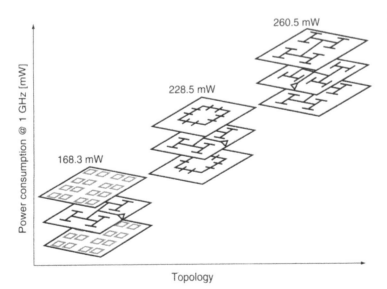

■ **FIGURE 10-28** Measured power consumption at 1 GHz of the different circuit blocks.

on the last level of the H-tree on the second plane. In addition, this topology requires a small amount of local interconnect resources as compared to the H-tree and global ring topologies. Most of the registers are connected directly to the local rings. Alternatively, the power consumed by the H-tree topology is the highest, as this topology requires three H-trees and additional wiring for the local connections to the leaves of each tree. In addition, the greatest number of buffers is included in this topology. This number is threefold as compared to the number of buffers used for the local mesh topology. Finally, the global rings block consumes slightly less power than the H-tree topology due to the reduced amount of wiring resources used by the global clock network.

Although the local mesh topology requires the least interconnect resources, a large number of 3-D vias is required for the interplane connections. Since the 3-D vias block all of the metal layers and occupy silicon area, the routing blockage increases considerably as compared to the H-tree topology. The global rings topology requires a moderate number of 3-D vias as only four connections between the vertices of the rings and the branch points of the H-tree are necessary.

Since three-dimensional integration greatly increases the complexity of designing a synchronization system, a topology that offers low overhead in the design process of a 3-D clock distribution network is preferable. From this perspective, a potential advantage of the H-tree topology is that each plane can be individually analyzed. This behavior occurs in an H-tree topology since the clock distribution network in each plane is exclusively connected to registers within the same plane. Alternatively, in the local ring topology, all of the registers from all of the planes, which are connected to each sink of the tree on the second plane, need to be simultaneously considered.

10.5 **SUMMARY**

A case study for investigating several clock distribution networks for 3-D ICs is described in this chapter, and measurements from a 3-D test circuit are presented. The characteristics of the circuit and related topologies are:

- A 3-D clock distribution network cannot be directly extended from a 2-D circuit due to the lack of symmetry in a 3-D circuit caused by the interplane vias.

- The 3-D FDSOI fabrication technology from MITLL has been used to manufacture the test circuit.

- The 3-D test circuit is composed of four independent blocks, where each block is a three-plane 3-D circuit. For each block, a different clock distribution network is utilized.

- All of the blocks in each plane share the same logic circuitry to emulate a variety of switching patterns in a synchronous digital circuit.

- The maximum clock frequency of the fabricated 3-D test circuit is 1.4 GHz.

- A comparison of the clock skew and power consumption of each block is provided. A topology combining the symmetry of an H-tree on the second plane and local meshes on the other two planes results in low clock skew while consuming the lowest power as compared to the other investigated 3-D topologies.

11

Conclusions

With MOSFET channel lengths several tens of nanometers long, modern silicon integrated systems can support gigascale transistor densities. On-chip systems, for example, often include several processing cores within a microprocessor system or a mixture of various silicon technologies, such as analog, digital, and RF, in a mixed-signal SoC. The former is the heart of a high-performance computing system, while the latter can feature a versatile, compact, and portable electronic product. A fundamental requirement for this system-on-chip paradigm is efficient and reliable communication among the system components.

High-speed, low-power, and low-noise intercomponent communication is fundamentally limited by the increasing impedance characteristics and length of the interconnect. Three-dimensional integration is a revolutionary solution to the deleterious effects of the long interconnects. In vertically integrated systems, the long interconnects spanning several millimeters are replaced by orders of magnitude shorter vertical wires. This inherent reduction in wirelength provides opportunities for increased speed, enhanced noise margins, and lower power.

Three-dimensional integrated systems can also comprise a variety of different silicon technologies, such as digital, analog, RF CMOS and SOI circuits, and nonsilicon semiconductor technologies, such as SiGe and GaAs, making 3-D systems highly suitable for a broad spectrum of applications. This salient characteristic raises a greater interest in three-dimensional integration as compared to other incremental solutions that simply scale CMOS technologies.

Within the context of an electronic product, circuit-level enhancements can profoundly affect fundamental system requirements, such as compactness, portability, and computing power. Since the application space for these systems is broad, a plethora of 3-D technologies are under development. These approaches range from monolithic 3-D circuits to polylithic and heterogeneous multiplane systems. The fundamental element of a 3-D silicon system can be a single transistor, a functional block, a bare die, or a packaged circuit. The structural granularity of the system, in turn, determines the form of vertical interconnections, which link the elements of a 3-D circuit in the vertical direction. The cost of manufacturing is directly related to the processing capabilities offered by these technologies. Fabrication processes that utilize TSVs for vertical interconnects are an economic form of vertical integration, while exhibiting considerable improvement in system performance.

A major outcome of this book is that effective solutions to physical design issues for 3-D circuits are demonstrated, emphasizing the significance of the TSVs. In the 3-D design process, the through silicon vias should be treated as a means to improve signaling among the physical planes within a 3-D stack.

The TSVs in a 3-D circuit provide synchronization, power and ground, and signaling among the planes and — in contrast to mainstream 2-D circuits — thermal cooling. The primary challenges and novel solutions for satisfying these objectives have been presented throughout the book.

The electrical characteristics of the TSVs affect all but the thermal objective. Consequently, the impedance characteristics of the TSV structures are considered in design algorithms and techniques to improve signaling. These methods enhance the benefits provided by the shorter interconnect length. The effectiveness of physical design techniques is considerably improved by exploiting the electrical behavior of these vias. This behavior is investigated in the case study of a fabricated 3-D circuit that evaluates the important issue of synchronization; a crucial objective in global signaling. Different ways to efficiently distribute the clock signal in the gigahertz regime are explored temporally and within three spatial dimensions for the first time. In a nutshell, enhanced interplane signaling is a prerequisite for high-performance and/or high-bandwidth 3-D computing applications.

High-performance applications typically include communications-limited architectures, such as a processor-memory system. Communication fabrics, such as on-chip networks and FPGAs, also greatly benefit from the short vertical interconnects. Different architectural configurations are explored in this book, demonstrating the many possible power and latency tradeoffs that result from exploiting the third dimension. Appropriate latency and power models supporting this analysis are also presented.

Finally, a systematic approach is provided for inserting and distributing the vertical interconnects to improve the overall thermal conductivity of a 3-D stack. By including the thermal objective in 3-D physical design algorithms and techniques, thermal gradients and high temperatures, which can degrade the reliability and performance of a 3-D circuit, are significantly reduced. Due to the high power densities, the thermal objective is an integral part of the physical design process for 3-D integrated systems. Presently, the lack of advanced packaging has shifted the focus of removing heat to primarily on-chip solutions. Design methodologies for circuit and package thermal co-design will offer superior solutions due to the high temperatures within 3-D circuits.

The major achievement of this book is the thorough exploration of multiple aspects of three-dimensional integration from manufacturing to physical design and algorithms to system level architectures. This comprehensive approach targets the design and analysis of 3-D integration from a circuits perspective. With this point of view, the primary objective in the development of this book is to provide intuition behind the most sensitive issues regarding the vertical interconnect, which is essentially intertwined within each step of the 3-D design process. The material described in this book is intended to shed light on those areas related to the design of 3-D integrated systems in an effort to develop large-scale multifunctional, multiplane systems to continue the microelectronics revolution.

Enumeration of Gate Pairs in a 3-D IC

The starting gates N_{start} used in the derivation of the interconnect length distribution for 3-D circuits in [123] are described in this appendix. Starting gates are these gates that can form manhattan hemispheres of radius l, where l is the interconnect length connecting two gates. For a 3-D circuit consisting of n planes, $N_n = N/n$ gates per plane. For those gates located close to the periphery of each plane, only partial manhattan hemispheres are formed. Consequently, the number of gates encircled by these partial manhattan hemispheres varies with the interconnect length:

Region I. $l = 0$

$$N_{start} = N_n n, \tag{A-1}$$

Region II. $0 < l \le (n-1)d_v$

$$N_{start} = n(N_n - l), \tag{A-2}$$

Region III. $(n-1)d_v < l \le \dfrac{\sqrt{N_n}}{2} + 1$

$$N_{start} = nN - l, \tag{A-3}$$

Region IV. $\dfrac{\sqrt{N_n}}{2} + 1 < l \le \sqrt{N_n}$,

$$N_{start} = nN_n - l - \left(l - \dfrac{\sqrt{N_n}}{2} - 1\right)\left(l - \dfrac{\sqrt{N_n}}{2}\right), \tag{A-4}$$

Region V. $\sqrt{N_n} < l \le \sqrt{N_n} + (n-1)d_v$

$$N_{start} = \dfrac{7N_n}{4} - \dfrac{\sqrt{N_n}}{2} - \sqrt{N_n}\,l - f\left[l, g\left[l - \sqrt{N_n} - 1,\right], 1\right], \tag{A-5}$$

Region VI. $\sqrt{N_n} + (n-1)d_v < l \le \dfrac{3\sqrt{N_n}}{2}$

$$N_{start} = \frac{7N_n}{4} - \frac{\sqrt{N_n}}{2} - \sqrt{N_n}l + f[l, n-1, 2] - 2f\left[l, n-1, \frac{3}{2}\right], \qquad \text{(A-6)}$$

Region VII. $\dfrac{3\sqrt{N_n}}{2} < l \le \dfrac{3\sqrt{N_n}}{2} + (n-1)d_v$

$$N_{start} = \left\{ \begin{array}{l} \left(2\sqrt{N_n} - l\right)\left(2\sqrt{N_n} - l - 1\right) + f[l, n-1, 2] - 2f\left[l, n-1, \dfrac{3}{2}\right] \\[4mm] +2f\left[l, g\left[l - \dfrac{3}{2}\sqrt{N_n} - 1, d_v\right], \dfrac{3}{2}\right] \end{array} \right\}, \qquad \text{(A-7)}$$

Region VIII. $\dfrac{3\sqrt{N_n}}{2} + (n-1)d_v < l \le 2\sqrt{N_n}$

$$N_{start} = \left(2\sqrt{N_n} - l\right)\left(2\sqrt{N_n} - l - 1\right) + f[l, n-1, 2], \qquad \text{(A-8)}$$

Region IX. $2\sqrt{N_n} < l \le 2\sqrt{N_n} + (n-1)d_v$

$$N_{start} = f[l, n-1, 2] - f\left[l, g\left[l - 2\sqrt{N_n} - 1, d_v\right], 2\right]. \qquad \text{(A-9)}$$

The function $g[x, y]$ is the discrete quotient function while function $f[x, y, z]$ is

$$f[l, y, z] = 2\left\{ \begin{array}{l} \dfrac{d_v^2 y}{6}\left(2y^2 + 3y + 1\right) + \dfrac{y}{2}\left(2z\sqrt{N_n}d_v - 2d_v l - d_v\right)(y+1) \\[4mm] + 4y\left(9N_n - 3\sqrt{N_n}l - \dfrac{3}{2}\sqrt{N_n} + l + l^2\right) \end{array} \right\}. \qquad \text{(A-10)}$$

d_v denotes the interplane distance between two adjacent planes in a 3-D system.

Formal Proof of Optimum Single Via Placement

In this appendix, the Lemma used to determine the optimum via location for an interplane interconnect that includes only one via for different values of r_{21}, c_{12}, and l_1 is stated and proved.

Lemma 1: If $f(x) = Ax^2 + Bx + C$ and $\dfrac{d^2f(x)}{dx^2} < 0$,

(a) for $x_{max} \in [x_0, x_1]$, $x_1 > x_0 > 0$,

(i) if $x_{max} > \dfrac{x_1 + x_0}{2}$, $f(x_0) < f(x_1)$, (ii) if $x_{max} < \dfrac{x_1 + x_0}{2}$, $f(x_1) < f(x_0)$,

(b) for $x_{max} < x_0$, $f(x_0) > f(x_1)$,

(c) for $x_{max} > x_1$, $f(x_0) < f(x_1)$.

Proof: (a) (i) $f(x)$ is a parabola with a symmetry axis at $x = x_{max}$. Thus, $f(x_{max} - x) = f(x_{max} + x)$. For $x = x_{max} - x_0$, $f(x_0) = f(2x_{max} - x_0)$. Since $\dfrac{d^2f(x)}{dl_1^2} < 0$, $f(x)$ is decreasing for $x > x_{max}$. From the hypothesis,

$x_{max} > \dfrac{x_1 + x_0}{2} \Leftrightarrow 2x_{max} - x_0 > x_1 > x_{max} \Leftrightarrow f(2x_{max} - x_0) < f(x_1)$, and since $f(x_0) = f(2x_{max} - x_0)$, $f(x_0) < f(x_1)$.

(ii) $f(x_0) = f(2x_{max} - x_0)$, and $2x_{max} - x_0 > x_{max}$ since from the hypothesis $x_{max} \in [x_0, x_1]$. By similar reasoning as in (i),

$x_{max} < \dfrac{x_1 + x_0}{2} \Leftrightarrow x_1 > 2x_{max} - x_0 \Leftrightarrow f(x_1) < f(2x_{max} - x_0) \Leftrightarrow f(x_1) < f(x_0)$.

(b) $f(x)$ decreases for $x > x_{max}$. Since $x_{max} < x_0$, for $x_{max} < x_0 < x_1$, $f(x_{max}) > f(x_0) > f(x_1)$.

(c) $f(x)$ increases for $x < x_{max}$. Since $x_{max} > x_0$, for $x_0 < x_1 < x_{max}$, $f(x_0) < f(x_1) < f(x_{max})$.

Proof of the Two-Terminal Via Placement Heuristic

A formal proof of the two-terminal heuristic for placing interplane vias is described in this appendix. Consider the following expression that describes the critical point (*i.e.*, the derivative of the delay is set equal to zero) for placing a via v_j, as illustrated in Fig. C-1,

$$x_j^* = -\left[\frac{l_{vj}\left(r_j c_{vj} - r_{vj}c_{j+1} + r_{j+1}c_{j+1} - r_j c_{j+1}\right) + R_{uj}\left(c_j - c_{j+1}\right) + \Delta x_j\left(r_j - r_{j+1}\right)c_{j+1} + C_{dj}\left(r_j - r_{j+1}\right)}{r_j c_j - 2r_j c_{j+1} + r_{j+1}c_{j+1}}\right].$$

$$(C\text{-}1)$$

From this expression, the critical point x_j is a monotonic function of the upstream resistance and downstream capacitance of the allowed interval for via v_j, denoted as R_{uj} and C_{dj}, respectively,

$$x_j^* = f\left(R_{vj}^*, C_{dj}^*\right). \qquad (C\text{-}2)$$

These quantities $(R_{uj}^*$ and $C_{dj}^*)$ depend upon the location of the other vias along the net and are unknown. However, as the allowed intervals for the vias and the impedance characteristics of the line are known, the minimum and maximum values of these impedances, $R_{uj\,\min}$, $R_{uj\,\max}$, $C_{dj\,\min}$, and $C_{dj\,\max}$, can be determined. Without loss of generality, assume that $r_j > r_{j+1}$ and $c_j > c_{j+1}$ (the other cases are similarly treated). For this case, the critical point (*i.e.*, $\frac{\partial T}{\partial x_j} = 0$) is a strictly increasing function of R_{uj} and C_{dj}. Consequently, the minimum and maximum value for the critical point $x_{j\,\min}^*$ and $x_{j\,\max}^*$ is determined from, respectively,

$$x_{j\,\min}^* = f\left(R_{uj\,\min}, C_{dj\,\min}\right), \qquad (C\text{-}3)$$

$$x_{j\,\max}^* = f\left(R_{uj\,\max}, C_{dj\,\max}\right). \qquad (C\text{-}4)$$

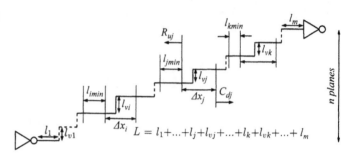

■ FIGURE C-1 Interplane interconnect consisting of m segments connecting two circuits located n planes apart.

The final value of the upstream (downstream) capacitance for via v_j, which is determined after placing all of the remaining vias of the net denoted as R_{uj}^* (C_{dj}^*) within the range, is

$$R_{uj\,min} < R_{uj}^* < R_{uj\,max}, \left(C_{dj\,min} < C_{dj}^* < C_{dj\,max}\right). \qquad \text{(C-5)}$$

Due to the monotonic relationship of the critical point x_j on R_{uj} and C_{dj},

$$x_{j\,min}^* = f\left(R_{uj\,min}, C_{dj\,min}\right) < x_j^* = f\left(R_{uj}^*, C_{dj}^*\right) < x_{j\,max}^* = f\left(R_{uj\,max}, C_{dj\,max}\right). \qquad \text{(C-6)}$$

Consequently, by iteratively decreasing the range of the x_j^* according to (C-6), the location for v_j can be determined.

To better explain this iterative procedure, an example is offered in Chapter 7 where the vias, v_i, v_j, and v_k, shown in Fig. C-1, have not yet been placed. In this example, vias v_i and v_k are assumed to belong to case *(iii)* of the heuristic. Since the allowed intervals for vias v_i, v_j, and v_k and the impedance characteristics of the respective horizontal segments are known, the minimum x_{min}^{*0} and maximum x_{max}^{*0} critical point for all of the segments i, j, and k are obtained. The minimum and maximum values of R_{ui}^0, R_{uj}^0, R_{uk}^0, C_{di}^0, C_{dj}^0, and C_{dk}^0 are determined, where the superscript represents the number of iterations.

From (C-6), the via location of segments i and k is contained within the limits determined by (C-1). As the interval for placing the vias v_i and v_k decreases, the minimum (maximum) value of the upstream resistance and downstream capacitance of segment j increases (decreases), *i.e.*, $R_{uj\,min}^0 < R_{uj\,min}^1$, $C_{dj\,min}^0 < C_{dj\,min}^1$, $R_{uj\,max}^1 < R_{uj\,max}^0$, and $C_{dj\,max}^1 < C_{dj\,max}^0$. Due to the monotonicity of x_j^* (see (C-2)-(C-4) and

(C-6)) on R_{uj} and C_{dj}, $x_{j\,min}^{*0} < x_{j\,min}^{*1}$ and $x_{j\,max}^{*1} < x_{j\,max}^{*0}$. The range of values for x_j^{*} therefore also decreases and, typically, after two or three iterations, the optimum location for the corresponding via is determined.

The above example is extended to each of the other possible cases that can occur for segments i and k. Specifically,

(a) i and k belong to either case (*i*) or (*ii*). Both R_{uj} and C_{dj} are precisely determined or, equivalently, $R_{uj\,min} = R_{uj\,max}$ and $C_{dj\,min} = C_{djmax}$. Consequently, the placement of both vias v_i and v_k is known and $x_{j\,min}^{*0} = x_{j\,max}^{*0} = x_j^{*}$ The placement of v_j is also determined within the first iteration.

(b) i belongs to case (*i*) or (*ii*) and k belongs to case (*iii*). R_{uj} is precisely determined or, equivalently, $R_{uj\,min} = R_{uj\,max}$ and the placement of via v_i is known. Since v_i is placed and k belongs to case (*iii*), $C_{dj\,min}^{0} < C_{dj\,min}^{1}$ and $C_{dj\,max}^{1} < C_{dj\,max}^{0}$. The placement of via v_j converges faster, as only the placement of segment k remains unknown after the first iteration.

(c) k belongs to case (i) or (ii) and i belongs to case (iii). C_{dj} is precisely determined or, equivalently, $C_{dj\,min} = C_{dj\,max}$ and the placement of via v_k is known. Since v_k is placed and i belongs to case (iii), $R_{uj\,min}^{0} < R_{uj\,min}^{1}$ and $R_{uj\,max}^{1} < R_{uj\,max}^{0}$. The placement of via v_j converges faster as only the placement of segment i remains unknown after the first iteration.

(d) i belongs to any of the cases (i)-(iii) and k belongs to case (iv). R_{uj} is readily determined (cases (i) and (ii)) or converges, as described in the previous example, $R_{uj\,min}^{0} < R_{uj\,min}^{1}$ and $R_{uj\,max}^{1} < R_{uj\,max}^{0}$ As k belongs to case (iv), however, C_{dj} does not change as in the cases above. If the decrease in the upstream resistance is sufficient to determine x_j^{*} according to (C-1), v_j is marked as processed, otherwise v_j is marked as unprocessed and the algorithm continues to the next via. In the latter case, the placement approach is described by case (iv) of the heuristic.

(e) k belongs to any of the cases (i)-(iii) and i belongs to case (iv). C_{dj} is readily determined (cases (i) and (ii)) or converges, as described in the previous example, implying $C_{dj\,min}^{0} < C_{dj\,min}^{1}$ and $C_{dj\,max}^{1} < C_{dj\,max}^{0}$. As i belongs to case (iv), however, R_{uj} does not change as in the aforementioned cases. Overall, if the decrease in the downstream capacitance is sufficient to determine x_j^{*}

according to (C-1), v_j is marked as processed, otherwise v_j is marked as unprocessed and the algorithm continues to the next via. In the latter case, the placement approach is described by case *(iv)* of the heuristic.

(f) Both i and k belong to case *(iv)*. Therefore, both R_{uj} and C_{dj} cannot be bounded. Consequently, v_j is marked as unprocessed and the next via is processed. Alternatively, this sub-case degenerates into case *(iv)* of the heuristic presented in Chapter 7.

Proof of Condition for Via Placement
of Multiterminal Nets

In this appendix, a proof for necessary *condition 1* is provided.

Condition 1: If $r_j > r_j{+}1$, only a *type-1* move for v_j can reduce the delay of a tree.

Proof: Consider Fig D-1 where the interplane via v_j (the solid square) can be placed in any direction d_e, d_s, and d_n within the interval l_{de}, l_{ds}, and l_{dn}, respectively. For the tree shown in Fig. D-1 and removing the terms that are independent of v_j, (8-1) is

$$T_w = \sum_{v_i \in U_{0j}} \sum_{s_p \in P_{s_p U_{ij}}} w_{s_p} R_{uij}\left(c_{v_j}l_{v_j} + C_{d_j}\right)$$

$$+ \sum_{s_p \in P_{s_p v_j}} w_{s_p}\left(R_{u_j}\left(c_{v_j}l_{v_j} + C_{d_j}\right) + r_{v_j}l_{v_j}C_{d_j} + \frac{r_{v_j}c_{v_j}l_{vj}^2}{2}\right), \tag{D-1}$$

where

$$C_{d_j} = \sum_{\forall k} C_{dv_j d_k} + c_{j+1}\left(l_{d_e} + l_{d_s} + l_{d_n}\right). \tag{D-2}$$

Suppose that a *type-2* move is required, shifting v_j by x towards the d_e direction (the dashed square). Expression (8-1) becomes

$$T'_w = \left[\begin{array}{l} \left(\displaystyle\sum_{v_i \in U_{ij}} \sum_{s_p \in P_{s_p U_{ij}}} w_{s_p} R_{uij} + \sum_{s_p \in P_{s_p v_j}} w_{s_p} R_{u_j}\right)\left(c_{v_j}l_{v_j} + c_j x + C_{d_j}\right) \\[2ex] + \displaystyle\sum_{s_p \in P_{s_p v_j}} w_{s_p}\left[\left(r_j - r_{j+1}\right)xC_{d_j} + r_{j+1}l_{d_e}\left(C_{d_j} - 1/2c_{j+1}l_{d_e}\right)\right. \\[2ex] \left. + r_j x\left(c_{v_j}l_{v_j} + C_{d_j}\right) + 1/2\left(r_{v_j}c_{v_j}l_{vj}^2 + r_j c_j x^2\right)\right] \end{array} \right]. \tag{D-3}$$

■ FIGURE D-1 A portion of an interconnect tree.

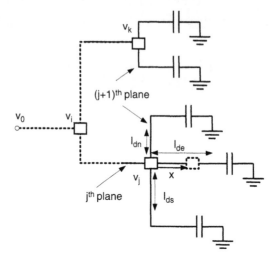

For a *type-2* move to reduce the weighted delay of the tree, shifting v_j should decrease T_w, or, equivalently, $\Delta T = T'_w - T_w < 0$. Subtracting (D-1) from (D-3) yields

$$\Delta T = \left\{ \sum_{s_p \in P_{s_p v_j}} w_{s_p} \left[r_j x \left(c_{v_j} l_{v_j} + C_{d_j} \right) + R_{u_j} c_j x + r_{j+1} l_{d_e} \left(C_{d_j} - \frac{c_{j+1} l_{d_e}}{2} \right) \right. \right. \\ \left. \left. + \left(r_j - r_{j+1} \right) x C_{d_j} + \frac{r_j c_j x_j^2}{2} \right] + \sum_{v_i \in U_{ij}} \sum_{s_p \in P_{s_p U_{ij}}} w_{s_p} R_{uij} c_j x \right\}.$$

(D-4)

Since $r_j > r_{j+1}$, and $C_{dj} > \frac{c_{j+1} l_{d_e}}{2}$ from (D-2), (D-4) is always positive and a *type-2* move cannot reduce the delay of a tree.

References

[1] C. Weiner, "How the Transistor Emerged," *IEEE Spectrum*, pp. 24-33, January 1977.

[2] P. H. Abelson and A. L. Hammond, "The Electronics Revolution," *Science*, Vol. 195, No. 4283, pp. 1087-1091, March 1977.

[3] N. P. Ruzic, "The Automated Factory: A Dream Come True?," *Control Engineering*, Vol. 25, No. 4, pp. 58-62, April 1978.

[4] J. C. Burns, "The Evolution of Office Information Systems," *Datamation*, pp. 60-64, April 1977.

[5] A. J. Nichols, "An Overview of Microprocessor Applications," *Proceedings of the IEEE*, Vol. 64, No. 6, pp. 951-953, June 1976.

[6] T. Forester, *The Microelectronics Revolution*, The MIT Press, Cambridge, Massachusetts, 1981.

[7] R. N. Noyce, "Microelectronics," *Scientific American*, Vol. 237, No. 3, pp. 62-69, September 1977.

[8] G. Lapidus, "Transistor Family History," *IEEE Spectrum*, pp. 34-35, January 1977.

[9] S. Augarten, *State of the Art: A Photographic History of the Integrated Circuit*, Ticknor & Fields, New Haven and New York, 1983.

[10] R. N. Noyce, "Large Scale Integration: What is Yet to Come," *Science*, Vol. 195, No. 4283, pp. 1102-1106, March 1977.

[11] K. C. Saraswat and F. Mohammadi, "Effect of Scaling of Interconnections on the Time Delay of VLSI Circuits," *IEEE Transactions on Electron Devices*, Vol. ED-29, No. 4, pp. 645-650, April 1982.

[12] *International Technology Roadmap for Semiconductors* ITRS, 2005 Edition.

[13] C. Akrout et al., "A 480-MHz RISC Microprocessor in a 0.12-µm L_{eff} CMOS Technology with Copper Interconnects," *IEEE Journal of Solid-State Circuits*, Vol. 33, No. 11, pp. 1609-1616, November 1998.

[14] D. H. Allen et al., "A 0.2 µm 1.8 V SOI 550 MHz 64 b Power PC Microprocessor with Copper Interconnects," *Proceedings of the IEEE International Solid-State Circuits Conference*, pp. 438-439, February 1999.

[15] M. Naik et al., "Process Integration of Double Level Copper-Low k (k=2.8) Interconnect," *Proceedings of the IEEE International Interconnect Technology Conference*, pp. 181-183, May 1999.

[16] P. Zarkesh-Ha et al., "The Impact of Cu/Low k on Chip Performance," *Proceedings of the IEEE International ASIC/SoC Conference*, pp. 257-261, September 1999.

[17] Y. Takao et al., "A 0.11 µm Technology with Copper and Very-Low-k Interconnects for High Performance System-on-Chip Cores," *Proceedings of the IEEE International Electron Device Meeting*, pp. 559-562, December 2000.

[18] J. D. Meindl, "Interconnect Opportunities for Gigascale Integration," *IEEE Micro*, Vol. 23, No. 3, pp. 28-35, May/June 2003.

[19] R. Venkatesan, J. A. Davis, K. A. Bowman, and J. D. Meindl, "Optimal n-tier Interconnect Architectures for Gigascale Integration (GSI)," *IEEE Transactions on Very Large Integration (VLSI) Systems*, Vol. 9, No. 6, pp. 899-912, December 2001.

[20] K. M. Lepak, I. Luwandi, and L. He, "Simultaneous Shield Insertion and Net Ordering under Explicit RLC Noise Constraint," *Proceedings of the IEEE/ACM Design Automation Conference*, pp. 199-202, June 2001.

[21] P. Fishburn, "Shaping a VLSI Wire to Minimize Elmore Delay," *Proceedings of the IEEE European Design and Test Conference*, pp. 244-251, March 1997.

[22] M. A. El-Moursy and E. G. Friedman, "Exponentially Tapered H-Tree Clock Distribution Networks," *IEEE Transactions on Very Large Scale Integration (VLSI) Systems*, Vol. 13, No. 8, pp. 971-975, August 2005.

[23] H. B. Bakoglu and J. D. Meindl, "Optimal Interconnection Circuits for VLSI," *IEEE Transactions on Electron Devices*, Vol. ED-32, No. 5, pp. 903-909, May 1985.

[24] Y. I. Ismail, E. G. Friedman, and J. L. Neves, "Exploiting On-Chip Inductance in High Speed Clock Distribution Networks," *IEEE Transactions on Very Large Scale Integration (VLSI) Systems*, Vol. 9, No. 6, pp. 963-973, December 2001.

[25] M. Ghoneima et al., "Reducing the Effective Coupling Capacitance in Buses Using Threshold Voltage Adjustment Techniques," *IEEE Transactions on Circuits and Systems I: Fundamental Theory and Applications*, Vol. 53, No. 9, pp. 1928-1933, September 2006.

[26] M. R. Stan and W. P. Burleson, "Bus-Invert Coding for Low-Power I/O," *IEEE Transactions on Very Large Scale Integration (VLSI) Systems*, Vol. 3, No. 1, pp. 49-58, March 1998.

[27] R. Bashirullah, L. Wentai, and R. K. Cavin III, "Current-Mode Signaling in Deep Submicrometer Global Interconnects," *IEEE Transactions on Very Large Scale Integration (VLSI) Systems*, Vol. 11, No. 3, pp. 406-417, June 2003.

[28] V. V. Deodhar and J. A. Davis, "Optimization of Throughput Performance for Low-Power VLSI Interconnects," *IEEE Transactions on Very Large Scale Integration (VLSI) Systems*, Vol. 13, No. 3, pp. 308-318, March 2005.

[29] H. Zhang, V. George, and J. M. Rabaey, "Low-Swing On-Chip Signaling Techniques: Effectiveness and Robustness," *IEEE Transactions on Very Large Scale Integration (VLSI) Systems*, Vol. 8, No. 3, pp. 264-272, June 2000.

[30] L. Benini and G. De Micheli, "Networks on Chip: A New SoC Paradigm," *IEEE Computer*, Vol. 31, No. 1, pp. 70-78, January 2002.

[31] M. Haurylau *et al.*, "On-Chip Optical Interconnect Roadmap: Challenges and Critical Directions," *IEEE Journal of Selected Topics on Quantum Electronics*, Vol. 12, No. 6, pp. 1699-1705, November/December 2006.

[32] G. Chen *et al.*, "On-Chip Copper-Based vs. Optical Interconnects: Delay Uncertainty, Latency, Power, and Bandwidth Density Comparative Predictions," *Proceedings of the IEEE International Interconnect Technology Conference*, pp. 39-41, June 2006.

[33] G. Chen, H. Chen, M. Haurylau, N. A. Nelson, D. H. Albonesi, P. M. Fauchet, and E. G. Friedman, "Predictions of CMOS Compatible On-Chip Optical Interconnect," *Integration, the VLSI Journal*, Vol. 40, No. 4, pp. 434-446, July 2007.

[34] G. T. Goele *et al.*, "Vertical Single Gate CMOS Inverters on Laser-Processed Multilayer Substrates," *Proceedings of the IEEE International Electron Device Meetings*, Vol. 27, pp. 554-556, December 1981.

[35] J. F. Gibbons and K. F. Lee, "One-Gate-Wide CMOS Inverter on Laser-Recrystallized Polysilicon," *IEEE Electron Device Letters*, Vol. EDL-1, No. 6, pp. 117-118, June 1980.

[36] W. R. Davis *et al.*, "Demystifying 3D ICs: the Pros and Cons of Going Vertical," *IEEE Design and Test of Computers*, Vol. 22, No. 6, pp. 498-510, November/December 2005.

[37] J. W. Joyner, P. Zarkesh-Ha, J. A. Davis, and J. D. Meindl, "A Three-Dimensional Stochastic Wire-Length Distribution for Variable Separation of Strata," *Proceedings of the IEEE International Interconnect Technology Conference*, pp. 126-128, June 2000.

[38] M. Koyanagi *et al.*, "Future System-on-Silicon LSI Chips," *IEEE Micro*, Vol. 18, No. 4, pp. 17-22, July/August 1998.

[39] V. K. Jain, S. Bhanja, G. H. Chapman, and L. Doddannagari, "A Highly Reconfigurable Computing Array: DSP Plane of a 3D Heterogeneous SoC," *Proceedings of the IEEE International Systems on Chip Conference*, pp. 243-246, September 2005.

[40] V. H. Nguyen and P. Christie, "The Impact of Interstratal Interconnect Density on the Performance of Three-Dimensional Integrated Circuits," *Proceedings of the IEEE/ACM International Workshop on System Level Interconnect Prediction*, pp. 73-77, April 2005.

[41] I. Savidis and E. G. Friedman, "Electrical Modeling and Characterization of 3-D Vias," *Proceedings of the IEEE International Symposium on Circuits and Systems*, pp. 784-787, May 2008.

[42] M. Karnezos, "3-D Packaging: Where All Technologies Come Together," *Proceedings of the IEEE/SEMI International Electronics Manufacturing Technology Symposium*, pp. 64-67, July 2004.

[43] J. Miettinen, M. Mantysalo, K. Kaija, and E. O. Ristolainen, "System Design Issues for 3D System-in-Package (SiP)," *Proceedings of the IEEE International Electronic Components and Technology Conference*, pp. 610-615, June 2004.

[44] S. F. Al-Sarawi, D. Abbott, and P. D. Franzon, "A Review of 3-D Packaging Technology," *IEEE Transactions on Components, Packaging, and Manufacturing Technology – Part B*, Vol. 21, No. 1, pp. 2-14, February 1998.

[45] R. J. Gutmann *et al.*, "Three-Dimensional (3D) ICs: A Technology Platform for Integrated Systems and Opportunities for New Polymeric Adhesives," *Proceedings of the IEEE International Conference on Polymers and Adhesives in Microelectronics and Photonics*, pp. 173-180, October 2001.

[46] E. Culurciello and A. G. Andreou, "Capacitive Inter-Chip Data and Power Transfer for 3-D VLSI," *IEEE Transactions on Circuits and Systems II: Express Briefs*, Vol. 53, No. 12, pp. 1348-1352, December 2006.

[47] R. R. Tummala *et al.*, "The SOP for Miniaturized, Mixed-Signal Computing, Communication, and Consumer Systems of the Next Decade," *IEEE Transactions on Advanced Packaging*, Vol. 27, No. 2, pp. 250-267, May 2004.

[48] R. R. Tummala, "SOP: What is it and Why? A New Microsystem-Integration Technology Paradigm-Moore's Law for System Integration of Miniaturized Convergent Systems of the Next Decade," *IEEE Transactions on Advanced Packaging*, Vol. 27, No. 2, pp. 241-249, May 2004.

[49] V. Sundaram *et al.*, "Next-Generation Microvia and Global Wiring Technologies for SOP," *IEEE Transactions on Advanced Packaging*, Vol. 27, No. 2, pp. 315-325, May 2004.

[50] H. P. Hofstee, "Future Microprocessors and Off-Chip SOP Interconnect," *IEEE Transactions on Advanced Packaging*, Vol. 27, No. 2, pp. 301-303, May 2004.

[51] P. Garrou, "Future ICs Go Vertical," *Semiconductor International* [online], February 2005.

[52] E. Beyne, "The Rise of the 3rd Dimension for System Integration," *Proceedings of the IEEE International Interconnect Technology Conference*, pp. 1-5, June 2006.

[53] A. C. Fox III and M. Warren, "High-Density Electronic Package Comprising Stacked Sub-Modules which are Electrically Interconnected by Solder-Filled Vias," U.S. Patent 5,128,831, October 1991.

[54] I. Miyano *et al.*, "Fabrication and Thermal Analysis of 3-D Located LSI Packages," *Proceedings of the European Hybrid Microelectronics Conference*, pp. 184-191, June 1993.

[55] S. Stoukach *et al.*, "3D-SiP Integration for Autonomous Sensor Nodes," *Proceedings of the IEEE International Electronic Components and Technology Conference*, pp. 404-408, June 2006.

[56] N. Tamaka *et al.*, "Low-Cost Through-Hole Electrode Interconnection for 3D-SiP Using Room-Temperature Bonding," *Proceedings of the IEEE International Electronic Components and Technology Conference*, pp. 814-818, June 2006.

[57] W. J. Howell *et al.*, "Area Array Solder Interconnection Technology for the Three-Dimensional Silicon Cube," *Proceedings of the IEEE International Electronic Components and Technology Conference*, pp. 1174-1178, May 1995.

[58] K. Hatada, H. Fujimoto, T. Kawakita, and T. Ochi, "A New LSI Bonding Technology 'Micron Bump Bonding Assembly Technology'," *Proceedings of the IEEE International Electronic Manufacturing Technology Symposium*, pp. 23-27, October 1988.

[59] J.-C. Souriau, O. Lignier, M. Charrier, and G. Poupon, "Wafer Level of 3D System in Package for RF and Data Applications," *Proceedings of the IEEE International Electronic Components and Technology Conference*, pp. 356-361, June 2005.

[60] K. Tanida *et al.*, "Ultra-High-Density 3D Chip Stacking Technology," *Proceedings of the IEEE International Electronic Components and Technology Conference*, pp. 1084-1089, May 2003.

[61] J. A. Minahan, A. Pepe, R. Some, and M. Suer, "The 3D Stack in Short Form," *Proceedings of the IEEE International Electronic Components and Technology Conference*, pp. 340-344, May 1992.

[62] S. P. Larcombe, J. M. Stern, P. A. Ivey, and L. Seed, "Utilizing a Low Cost 3D Packaging Technology for Consumer Applications," *IEEE Transactions on Consumer Electronics*, Vol. 41, No. 4, pp. 1095-1102, November 1995.

[63] J. M. Stern *et al.*, "An Ultra Compact, Low-Cost, Complete Image-Processing System," *Proceedings of the IEEE International Solid-State Circuits Conference*, pp. 230-231, February 1995.

[64] J. Barrett *et al.*, "Performance and Reliability of a Three-Dimensional Plastic Moulded Vertical Multichip Module (MCM-V)," *Proceedings of the IEEE International Electronic Components and Technology Conference*, pp. 656-663, May 1995.

[65] C. C. Liu, J.-H. Chen, R. Manohar, and S. Tiwari, "Mapping System-on-Chip Designs from 2-D to 3-D ICs," *Proceedings of the IEEE International Symposium on Circuits and Systems*, pp. 2939-2942, May 2005.

[66] R. Weerasekera, L. R. Zheng, D. Pamunuwa, and H. Tenhunen, "Extending Systems-on-Chip to the Third Dimension: Performance, Cost and Technological Tradeoffs," *Proceedings of the IEEE/ACM International Conference on Computer-Aided Design*, pp. 212-219, November 2007.

[67] M. W. Geis, D. C. Flanders, D. A. Antoniadis, and H. I. Smith, "Crystalline Silicon on Insulators by Graphoepitaxy," *Proceedings of the IEEE International Electron Devices Meeting*, pp. 210-212, December 1979.

[68] S. Akiyama *et al.*, "Multilayer CMOS Device Fabricated on Laser Recrystallized Silicon Islands," *Proceedings of the IEEE International Electron Devices Meeting*, pp. 352-355, December 1983.

[69] S. Kawamura *et al.*, "Three-Dimensional CMOS IC's Fabricated by Using Beam Recrystallization," *IEEE Electron Device Letters*, Vol. EDL-4, No. 10, pp. 366-368, October 1983.

[70] K. Sugahara *et al.*, "SOI/SOI/Bulk-Si Triple-Level Structure for Three-Dimensional Devices," *IEEE Electron Device Letters*, Vol. EDL-7, No. 3, pp. 193-194, March 1986.

[71] K. F. Lee, J. F. Gibbons, and K. C. Saraswat, "Thin Film MOSFET's Fabricated in Laser-Annealed Polycrystalline Silicon," *Applied Physics Letters*, Vol. 35, No. 2, pp. 173-175, July 1979.

[72] H. Hazama *et al.*, "Application of E-beam Recrystallization to Three-Layer Image Processor Fabrication," *IEEE Transactions on Electron Devices*, Vol. 38, No. 1, pp. 47-54, January 1991.

[73] V. W. Chan, P. C. H. Chan, and M. Chan, "Three-Dimensional CMOS Integrated Circuits on Large Grain Polysilicon Films," *Proceedings of the IEEE International Electron Devices Meeting*, pp. 161-164, December 2000.

[74] V. Subramania and K. C. Saraswat, "High-Performance Germanium-Seeded Laterally Crystallized TFT's for Vertical Device Integration," *IEEE Transactions on Electron Devices*, Vol. 45, No. 9, pp. 1934-1939, September 1998.

[75] G. W. Neudeck, S. Pae, J. P. Denton, and T. C. Su, "Multiple Layers of Silicon-on-Insulator for Nanostructure Devices," *Journal of Vacuum Science Technology B*, Vol. 17, No. 3, pp. 994-998, May/June 1999.

[76] N. Hirashita, T. Katoh, and H. Onoda, "Si-Gate CMOS Devices on a Si Lateral Solid-Phase Epitaxial Layer," *IEEE Transactions on Electron Devices*, Vol. 36, No. 3, pp. 548-552, March 1989.

[77] X. Lin, S. Zhang, X. Wu, and M. Chan, "Local Clustering 3-D Stacked CMOS Technology for Interconnect Loading Reduction," *IEEE Transactions on Electron Devices*, Vol. 53, No. 6, pp. 1405-1410, June 2006.

[78] H.-S. P. Wong, K. K. Chan, and Y. Tuar, "Self-Aligned (Top and Bottom) Double-Gate MOSFET with a 25 nm Thick Silicon Channel," *Proceedings of the IEEE International Electron Devices Meeting*, pp. 427-430, December 1997.

[79] R. S. Shenoy and K. C. Saraswat, "Novel Process for Fully Self-Aligned Planar Ultrathin Body Double-Gate FET," *Proceedings of the IEEE International Silicon on Insulator Conference*, pp. 190-191, October 2004.

[80] B. Yu *et al.*, "FinFet Scaling to 10 nm Gate Length," *Proceedings of the IEEE International Electron Devices Meeting*, pp. 251-254, December 2002.

[81] X. Wu *et al.*, "A Three-Dimensional Stacked Fin-CMOS Technology for High-Density ULSI Circuits," *IEEE Transactions on Electron Devices*, Vol. 52, No. 9, pp. 1998-2003, September 2005.

[82] X. Wu *et al.*, "Stacked 3-D Fin-CMOS Technology," *IEEE Electron Device Letters*, Vol. 26, No. 6, pp. 416-418, June 2005.

[83] A. Fan, A. Rahman, and R. Reif, "Copper Wafer Bonding," *Electrochemical and Solid-State Letters*, Vol. 2, No. 10, pp. 534-536, October 1999.

[84] R. Reif, A. Fan, K. N. Chen, and S. Das, "Fabrication Technologies for Three-Dimensional Integrated Circuits," *Proceedings of the IEEE International Symposium on Quality Electronic Design*, pp. 33-37, March 2002.

[85] J.-Q. Lu *et al.*, "Stacked Chip-to-Chip Interconnections Using Wafer Bonding Technology with Dielectric Bonding Glues," *Proceedings of the IEEE International Interconnect Technology Conference*, pp. 219-221, June 2001.

[86] A. Klumpp, R. Merkel, R. Wieland, and P. Ramm, "Chip-to-Wafer Stacking Technology for 3D System Integration," *Proceedings of the IEEE International Electronic Components and Technology Conference*, pp. 1080-1083, May 2003.

[87] C. A. Bower *et al.*, "High Density Vertical Interconnect for 3-D Integration of Silicon Integrated Circuits," *Proceedings of the IEEE International Electronic Components and Technology Conference*, pp. 399-403, June 2006.

[88] T. Fukushima, Y. Yamada, H. Kikuchi, and M. Koyanagi, "New Three-Dimensional Integration Using Self-Assembly Technique," *Proceedings of the IEEE International Electron Devices Meeting*, pp. 348-351, December 2005.

[89] A. W. Topol *et al.*, "Enabling SOI-Based Assembly Technology for Three-Dimensional (3D) Integrated Circuits (ICs)," *Proceedings of the IEEE International Electron Devices Meeting*, pp. 352-355, December 2005.

[90] S. Tiwari *et al.*, "Three-Dimensional Integration for Silicon Electronics," *Proceedings of the IEEE Lester Eastman Conference on High Performance Devices*, pp. 24-33, August 2002.

[91] A. R. Mirza, "One Micron Precision, Wafer-Level Aligned Bonding for Interonnect, MEMS and

Packaging Applications," *Proceedings of the IEEE International Electronic Components and Technology Conference*, pp. 676-680, May 2000.

[92] N. Watanabe, T. Kojima, and T. Asano, "Wafer-Level Compliant Bump for Three-Dimensional LSI with High-Density Area Bump Connections," *Proceedings of the IEEE International Electron Devices Meeting*, pp. 671-674, December 2005.

[93] R. N. Vrtis, K. A. Heap, W. F. Burgoyne, and L. M. Roberson, "Poly (Arylene Ethers) as Low Dielectric Constant Materials for ULSI Interconnect Applications," *Proceedings of the Materials Research Society Symposium*, Vol. 443, p. 171, December 1997.

[94] N. H. Hendricks, "The Status of Low-k Materials Development," *Proceedings of the International VLSI Multilevel Interconnect Conference*, p. 17, June 2000.

[95] S. F. Hahn, S. J. Martin, and M. L. McKelvy, "Thermally Induced Polymerization of an Arylvinylbenzocyclobenzene Monomer," *Macromolecules*, Vol. 25, No. 5, pp. 1539-1545, September 1992.

[96] D. Oben, P. Weigand, M. J. Shapiro, and S. A. Cohen, "Influence of the Cure Process on the Properties of Hydrogen Silsequioxene Spin-on-Glass," *Proceedings of the Materials Research Society Symposium*, Vol. 443, p. 195, December 1997.

[97] T.-M. Lu and J. A. Moor, "Vapor Deposition of Low-Dielectric-Constant Polymeric Thin Films," *Materials Research Bulletin*, Vol. 22, No. 10, pp. 28-32, October 1997.

[98] S. A. Kühn, M. B. Kleiner, R. Thewes, and W. Weber, "Vertical Signal Transmission in Three-Dimensional Integrated Circuits by Capacitive Coupling," *Proceedings of the IEEE International Symposium on Circuits and Systems*, Vol. 1, pp. 37-40, May 1995.

[99] A. Fazzi et al., "3-D Capacitive Interconnections for Wafer-Level and Die-Level Assembly," *IEEE Journal of Solid-State Circuits*, Vol. 42, No. 10, pp. 2270-2282, October 2007.

[100] J. Xu et al., "AC Coupled Interconnect for Dense 3-D ICs," *IEEE Transactions on Nuclear Science*, Vol. 51, No. 5, pp. 2156-2160, October 2004.

[101] B. Kim et al., "Factors Affecting Copper Filling Process Within High Aspect Ratio Deep Vias for 3D Chip Stacking," *Proceedings of the IEEE International Electronic Components and Technology Conference*, pp. 838-843, June 2006.

[102] M. W. Newman et al., "Fabrication and Electrical Characterization of 3D Vertical Interconnects," *Proceedings of the IEEE International Electronic Components and Technology Conference*, pp. 394-398, June 2006.

[103] N. T. Nguyen et al., "Through-Wafer Copper Electroplating for Three-Dimensional Interconnects," *Journal of Micromechanics and Microengineering*, Vol. 12, No. 4, pp. 395-399, July 2002.

[104] C. S. Premachandran et al., "A Vertical Wafer Level Packaging Using Through Hole Filled Via Interconnect by Lift-Off Polymer Method for MEMS and 3D Stacking Applications," *Proceedings of the IEEE International Electronic Components and Technology Conference*, pp. 1094-1098, June 2005.

[105] F. Laermer, P. Schilp, and R. Bosch Gmbh, "Method of Anisotropically Etching Silicon," U.S. Patent 5,501,893, 1996; German Patent DE4241045C1, 1994.

[106] R. Nagarajan et al., "Development of a Novel Deep Silicon Tapered Via Etch Process for Through-Silicon Interconnection in 3-D Integrated Systems," *Proceedings of the IEEE International Electronic Components and Technology Conference*, pp. 383-387, June 2006.

[107] P. Dixit and J. Miao, "Fabrication of High Aspect Ratio 35 µm Pitch Interconnects for Next Generation 3-D Wafer Level Packaging by Through-Wafer Copper Electroplating," *Proceedings of the IEEE International Electronic Components and Technology Conference*, pp. 388-393, June 2006.

[108] S. X. Zhang, S.-W. R. Lee, L. T. Weng, and S. So, "Characterization of Copper-to-Silicon for the Application of 3D Packaging with Through Silicon Vias," *Proceedings of the IEEE International Conference on Electronic Packaging Technology*, pp. 51-56, September 2005.

[109] N. Ranganathan et al., "High Aspect Ratio Through-Wafer Interconnect for Three-Dimensional Integrated Circuits," *Proceedings of the IEEE International Electronic Components and Technology Conference*, pp. 343-348, June 2005.

[110] D. Henry et al., "Low Electrical Resistance Silicon Through Vias: Technology and Characterization," *Proceedings of the IEEE International Electronic Components and Technology Conference*, pp. 1360-1365, June 2006.

[111] C. Odoro *et al.*, "Analysis of the Induced Stresses in Silicon During Thermocompression Cu-Cu Bonding of Cu-Through-Vias in 3D-SIC Architecture," *Proceedings of the IEEE International Electronic Components and Technology Conference*, pp. 249-255, June 2007.

[112] D. Sabuncuoglu-Tezcan *et al.*, "Sloped Through Wafer Vias for 3D Wafer Level Packaging," *Proceedings of the IEEE International Electronic Components and Technology Conference*, pp. 643-647, June 2007.

[113] R. S. Patti, "Three-Dimensional Integrated Circuits and the Future of System-on-Chip Designs," *Proceedings of the IEEE*, Vol. 94, No. 6, pp. 1214-1224, June 2006.

[114] D. M. Jang *et al.*, "Development and Evaluation of 3-D SiP with Vertically Interconnected Through Silicon Vias (TSV)," *Proceedings of the IEEE International Electronic Components and Technology Conference*, pp. 847-850, June 2007.

[115] B. S. Landman and R. L. Russo, "On a Pin Versus Block Relationship for Partitions of Logic Graphs," *IEEE Transactions on Computers*, Vol. C-20, No. 12, pp. 1469-1479, December 1971.

[116] W. E. Donath, "Placement and Average Interconnection Lengths of Computer Logic," *IEEE Transactions on Circuits and Systems*, Vol. 26, No. 4, pp. 272-277, April 1979.

[117] P. Christie and D. Stroobandt, "The Interpretation and Application of Rent's Rule," *IEEE Transactions on Very Large Scale Integration (VLSI) Systems*, Vol. 8, No. 6, pp. 639-648, December 2000.

[118] P. Verplaetse, D. Stroobandt, and J. Van Campenhout, "A Stochastic Model for the Interconnection Topology of Digital Circuits," *IEEE Transactions on Very Large Scale Integration (VLSI) Systems*, Vol. 9, No. 6, pp. 938-942, December 2001.

[119] A. B. Kahng, S. Mantik, and D. Stroobandt, "Toward Accurate Models of Achievable Routing," *IEEE Transactions on Computer-Aided Design of Integrated Circuits and Systems*, Vol. 20, No. 5, pp. 648-659, May 2001.

[120] D. Stroobandt, *A Priori Wire Length Estimates for Digital Design*, Kluwer Academic Publishers, Netherlands 2001.

[121] J. A. Davis, V. K. De, and J. D. Meindl, "A Stochastic Wire-Length Distribution for Gigascale Integration (GSI) – Part I: Derivation and Validation," *IEEE Transactions on Electron Devices*, Vol. 45, No. 3, pp. 580-589, March 1998.

[122] J. A. Davis, V. K. De, and J. D. Meindl, "A Stochastic Wire-Length Distribution for Gigascale Integration (GSI) – Part II: Applications to Clock Frequency, Power Dissipation, and Chip Size Estimation," *IEEE Transactions on Electron Devices*, Vol. 45, No. 3, pp. 590-597, March 1998.

[123] J. W. Joyner *et al.*, "Impact of Three-Dimensional Architectures on Interconnects in Gigascale Integration," *IEEE Transactions on Very Large Scale Integration (VLSI) Systems*, Vol. 9, No. 6, pp. 922-928, December 2001.

[124] J. W. Joyner, P. Zarkesh-Ha, J. A. Davis, and J. D. Meindl, "Vertical Pitch Limitations on Performance Enhancement in Bonded Three-Dimensional Interconnect Architectures," *Proceedings of the ACM International System Level Interconnect Prediction Conference*, pp. 123-127, April 2000.

[125] J. W. Joyner, P. Zarkesh-Ha, and J. D. Meindl, "A Stochastic Global Net-Length Distribution for a Three-Dimensional System-on-a-Chip (3D-SoC)," *Proceedings of the IEEE International ASIC/SOC Conference*, pp. 147-151, September 2001.

[126] J. W. Joyner, *Opportunities and Limitations of Three-Dimensional Integration for Interconnect Design*, Ph. D. Dissertation, Georgia Institute of Technology, Atlanta, Georgia, July 2003.

[127] A. Rahman, A. Fan, J. Chung, and R. Reif, "Wire-Length Distribution of Three-Dimensional Integrated Circuits," *Proceedings of the IEEE International Interconnect Technology Conference*, pp. 233-235, May 1999.

[128] A. Rahman and R. Reif, "System Level Performance Evaluation of Three-Dimensional Integrated Circuits," *IEEE Transactions on Very Large Scale Integration (VLSI) Systems*, Vol. 8, No. 6, pp. 671-678, December 2000.

[129] A. Rahman, A. Fan, and R. Reif, "Comparison of Key Performance Metrics in Two- and Three-Dimensional Integrated Circuits," *Proceedings of the IEEE International Interconnect Technology Conference*, pp. 18-20, June 2000.

[130] R. Zhang, K. Roy, C.-K. Koh, and D. B. Janes, "Stochastic Interconnect Modeling, Power Trends, and Performance Characterization of 3-D Circuits," *IEEE Transactions on Electron Devices*, Vol. 48, No. 4, pp. 638-652, April 2001.

[131] R. Zhang, K. Roy, C.-K. Koh, and D. B. Janes, "Power Trends and Performance Characterization of 3-Dimensional Integration," *Proceedings of the IEEE International Symposium on Circuits and Systems*, Vol. IV, pp. 414-417, May 2001.

[132] D. Stroobandt and J. Van Campenhout, "Accurate Interconnection Lengths in Three-Dimensional Computer Systems," *IEICE Transactions on Information and Systems, Special Issue on Physical Design in Deep Submicron*, Vol. 10, No. 1, pp. 99-105, April 2000.

[133] D. Stroobandt, H. Van Marck, and J. Van Campenhout, "On the Use of Generating Polynomials for the Representation of Interconnection Length Distributions," *Proceedings of the International Workshop on Symbolic Methods and Applications to Circuit Design*, pp. 74-78, October 1996.

[134] D. Stroobandt, "Improving Donath's Technique for Estimating the Average Interconnection Length in Computer Logic," *Technical Report DG 96-01*, Ghent University, Belgium, ELIS Department, June 1996.

[135] W. E. Donath, "Wire Length Distribution for Placements of Computer Logic," *IBM Journal of Research and Development*, Vol. 25, No. 2/3, pp. 152-155, May 1981.

[136] K. C. Saraswat, S. K. Souri, K. Banerjee, and P. Kapour, "Performance Analysis and Technology of 3-D ICs," *Proceedings of the ACM International System Level Interconnect Prediction Conference*, pp. 85-90, April 2000.

[137] K. Banerjee, S. K. Souri, P. Kapour, and K. C. Saraswat, "3-D ICs: A Novel Chip Design Paradigm for Improving Deep-Submicrometer Interconnect Performance and Systems-on-Chip Integration," *Proceedings of the IEEE*, Vol. 89, No. 5, pp. 602-633, May 2001.

[138] J. W. Joyner, P. Zarkesh-Ha, and J. D. Meindl, "A Global Interconnect Design Window for a Three-Dimensional System-on-a-Chip," *Proceedings of the IEEE International Interconnect Technology Conference*, pp. 154-156, June 2001.

[139] J. W. Joyner and J. D. Meindl, "Opportunities for Reduced Power Distribution Using Three-Dimensional Integration," *Proceedings of the IEEE International Interconnect Technology Conference*, pp. 148-150, June 2002.

[140] "FDSOI Design Guide," MIT Lincoln Laboratories, Cambridge, 2006.

[141] H. Hua *et al.*, "Performance Trend in Three-Dimensional Integrated Circuits," *Proceedings of the IEEE International Interconnect Technology Conference*, pp. 45-47, June 2006.

[142] OpenRISC Reference Platform System-on-a-Chip and OpenRISC I200 IP Core Specification, online [http://www.opencores.org/projects.cgi/web/orlk/orpso].

[143] K. Bernstein *et al.*, "Interconnects in the Third Dimension: Design Challenges for 3-D ICs," *Proceedings of the IEEE/ACM Design Automation Conference*, pp. 562-567, June 2007.

[144] S. A. Kühn, M. B. Kleiner, P. Ramm, and W. Weber, "Performance Modeling of the Interconnect Structure of a Three-Dimensional Integrated RISC Processor/Cache System," *IEEE Transactions on Components, Packaging, and Manufacturing Technology – Part B*, Vol. 19, No. 4, pp. 719-727, November 1996.

[145] R. H. J. M. Otten, "Automatic Floorplan Design," *Proceedings of the IEEE/ACM Design Automation Conference*, pp. 261-267, June 1982.

[146] X. Hong *et al.*, "Corner Block List: An Effective and Efficient Topological Representation of Non-Slicing Floorplan," *Proceedings of the IEEE/ACM International Conference on Computer-Aided Design*, pp. 8-11, November 2000.

[147] E. F. Y. Yong, C. C. N. Chu, and C. S. Zion, "Twin Binary Sequences: A Non-Redundant Representation for General Non-Slicing Floorplan," *IEEE Transactions on Computer-Aided Design of Integrated Circuits and Systems*, Vol. 22, No. 4, pp. 457-469, April 2003.

[148] J. M. Lin and Y. W. Chang, "TCG: A Transitive Closure Graph Based Representation for Non-Slicing Floorplans," *Proceedings of the IEEE/ACM Design Automation Conference*, pp. 764-769, June 2001.

[149] H. Yamazaki, K. Sakanushi, S. Nakatake, and Y. Kajitani, "The 3D-Packing by Meta Data Structure and Packing Heuristics," *IEICE Transactions on Fundamentals of Electronics, Communications and Computer Sciences*, Vol. E83-A, No. 4, pp. 639-645, April 2000.

[150] L. Cheng, L. Deng, and D. F. Wong, "Floorplanning for 3-D VLSI Design," *Proceedings of the IEEE Asia and South Pacific Design Automation Conference*, pp. 405-411, January 2005.

[151] Z. Li *et al.*, "Hierarchical 3-D Floorplanning Algorithm for Wirelength Optimization," *IEEE Transactions on Circuits and Systems I: Regular Papers*, Vol. 53, No. 12, pp. 2637-2646, December 2006.

[152] Y. Deng and W. P. Maly, "Interconnect Characteristics of 2.5-D System Integration Scheme," *Proceedings of the IEEE International Symposium on Physical Design*, pp. 341-345, April 2001.

[153] S. Salewski and E. Barke, "An Upper Bound for 3D Slicing Floorplans," *Proceedings of the IEEE Asia and South Pacific Design Automation Conference*, pp. 567-572, January 2002.

[154] P. H. Shiu, R. Ravichandran, S. Easwar, and S. K. Lim, "Multi-Layer Floorplanning for Reliable System-on-Package," *Proceedings of the IEEE International Symposium on Circuits and Systems*, Vol. V, pp. 69-72, May 2004.

[155] J. Cong, J. Wei, and Y. Zhang, "A Thermal-Driven Floorplanning Algorithm for 3-D ICs," *Proceedings of the IEEE/ACM International Conference on Computer-Aided Design*, pp. 306-313, November 2004.

[156] T. Yan, Q. Dong, Y. Takashima, and Y. Kajitani, "How Does Partitioning Matter for 3D Floorplanning," *Proceedings of the ACM International Great Lakes Symposium on VLSI*, pp. 73-76, April/May 2006.

[157] [Online]. Available: http://www.cse.ucsc.edu/research/surf/GSRC/progress.html

[158] X. Hong *et al.*, "Non-Slicing Floorplan and Placement Using Corner Block List Topological Representation," *IEEE Transactions on Circuits and Systems II: Express Briefs*, Vol. 51, No. 5, pp. 228-233, May 2004.

[159] M. Healy *et al.*, "Multiobjective Microarchitectural Floorplanning for 2-D and 3-D ICs," *IEEE Transactions on Computer-Aided Design of Integrated Circuits and Systems*, Vol. 26, No. 1, pp. 38-52, January 2007.

[160] P. Shivakumar and N. P. Jouppi, "CACTI 3.0: An Integrated Cache Timing, Power, and Area Model," HP Western Research Labs, Palo Alto, CA, Technical Report 2001.2, 2001.

[161] J. C. Eble, V. K. De, D. S. Wills, and J. D. Meindl, "A Generic System Simulator (GENESYS) for ASIC Technology and Architecture Beyond 2001," *Proceedings of the IEEE International ASIC Conference*, pp. 193-196, September 1996.

[162] T. M. Austin, *Simplescalar Tool Suite*. [Online]. Available: http://www.simplescalar.com

[163] D. Brooks, V. Tiwari, and M. Martonosi, "Wattch: A Framework for Architectural-Level Power Analysis and Optimizations," *Proceedings of the IEEE International Conference on Computer Architecture*, pp. 83-94, June 2000.

[164] N. A. Sherwani, *Algorithms for VLSI Physical Design Automation*, Kluwer Academic Publishers, 3rd edition, 2002.

[165] M. Ohmura, "An Initial Placement Algorithm for 3-D VLSI," *Proceedings of the IEEE International Symposium on Circuits and Systems*, Vol. IV, pp. 195-198, May 1998.

[166] T. Tanprasert, "An Analytical 3-D Placement that Preserves Routing Space," *Proceedings of the IEEE International Symposium on Circuits and Systems*, Vol. III, pp. 69-72, May 2000.

[167] I. Kaya, M. Olbrich, and E. Barke, "3-D Placement Considering Vertical Interconnects," *Proceedings of the IEEE International SOC Conference*, pp. 257-258, September 2003.

[168] R. Hentschke and R. Reis, "A 3D-Via Legalization Algorithm for 3D VLSI Circuits and its Impact on Wire Length," *Proceedings of the IEEE International Symposium on Circuits and Systems*, pp. 2036-2039, May 2007.

[169] R. Hentschke, G. Flach, F. Pinto, and R. Reis, "Quadratic Placement for 3D Circuits Using Z-Cell Shifting, 3D Iterative Refinement and Simulated Annealing," *Proceedings of the ACM International Symposium on Integrated Circuits and System Design*, pp. 220-225, September 2006.

[170] E. Wong, J. Minz, and S. K. Lim, "Multi-Objective Module Placement for 3-D System-On-Package," *IEEE Transactions on Very Large Scale Integration (VLSI) Systems*, Vol. 14, No. 5, pp. 553-557, May 2006.

[171] H. Murata, K. Fujiyoshi, S. Nakatake, and Y. Kajitani, "Rectangle Packing Based Module Placement," *Proceedings of the IEEE International Conference on Computer-Aided Design*, pp. 472-479, November 1995.

[172] E. Wong, J. Minz, and S. K. Lim, "Power Supply Noise-Aware 3D Floorplanning for System-on-Package," *Proceedings of the IEEE Topical Meeting on Electrical Performance on Electronic Packaging*, pp. 259-262, October 2005.

[173] M. Popovich and E. G. Friedman, "Decoupling Capacitors for Multi-Voltage Power Distribution Systems," *IEEE Transactions on Very Large Scale Integration (VLSI) Systems*, Vol. 14, No. 3, pp. 217-228, March 2006.

[174] A. Mezhiba and E. G. Friedman, *Power Distribution Networks in High Speed Integrated Circuits*, Kluwer Academic Publishers, 2004.

[175] M. Popovich, A. V. Mezhiba, and E. G. Friedman, *Power Distribution Networks with On-Chip Decoupling Capacitors*, Springer Verlag, 2008.

[176] J. Minz, S. K. Lim, and C.-K. Koh, "3D Module Placement for Congestion and Power Noise Reduction," *Proceedings of the ACM International Great Lakes Symposium on VLSI*, pp. 458-461, April 2005.

[177] [Online]. Available: http://www.gtcad.gatech.edu

[178] R. J. Enbody, G. Lynn, and K. H. Tan, "Routing the 3-D Chip," *Proceedings of the IEEE/ACM Design Automation Conference*, pp. 132-137, June 1991.

[179] S. Tayu and S. Ueno, "On the Complexity of Three-Dimensional Channel Routing," *Proceedings of the IEEE International Symposium on Circuits and Systems*, pp. 3399-3402, May 2007.

[180] C. C. Tong and C.-L. Wu, "Routing in a Three-Dimensional Chip," *IEEE Transactions on Computers*, Vol. 44, No. 1, pp. 106-117, January 1995.

[181] A. Hashimoto and J. Stevens, "Wire Routing by Optimizing Channel Assignment within Large Apertures," *Proceedings of the IEEE/ACM Design Automation Conference*, pp. 155-169, June 1971.

[182] T. Ohtsuki, *Advances in CAD for VLSI: Vol. 4, Layout Design and Verification*, Elsevier, 1986.

[183] J. Minz and S. K. Lim, "Block-Level 3-D Global Routing With an Application to 3-D Packaging," *IEEE Transactions on Computer-Aided Design of Integrated Circuits and Systems*, Vol. 25, No. 10, pp. 2248-2257, October 2006.

[184] A. Harter, *Three-Dimensional Integrated Circuit Layout*, Cambridge University Press, 1991.

[185] B. Hoefflinger, S. T. Liu, and B. Vajdic, "A Three-Dimensional CMOS Design Methodology," *IEEE Transactions on Electron Devices*, Vol. ED-31, No. 2, pp. 171-173, February 1984.

[186] S. M. Alam, D. E. Troxel, and C. V. Thompson, "A Comprehesive Layout Methodology and Layout-Specific Circuit Analyses for Three-Dimensional Integrated Circuits,"

Proceedings of the IEEE International Symposium on Quality Electronic Design, pp. 246-251, March 2002.

[187] S. Das, A. Chandrakasan, and R. Reif, "Design Tools for 3-D Integrated Circuits," *Proceedings of the IEEE Asia and South Pacific Design Automation Conference*, pp. 53-56, January 2003.

[188] [Online]. Available: http://www.ece.ncsu.edu/erl/3DIC/pub

[189] K. Banerjee, A. Mehrotra, A. Sangiovanni-Vincentelli, and C. Hu, "On Thermal Effects in Deep Sub-Micron VLSI Interconnects," *Proceedings of the IEEE/ACM Design Automation Conference*, pp. 885-890, June 1999.

[190] C. H. Tsai and S.-M. Kang, "Cell-Level Placement for Improving Substrate Thermal Distribution," Vol. 19, No. 2, pp. 253-266, February 2000.

[191] V. Szekely, M. Rencz, and B. Courtois, "Tracing the Thermal Behavior of ICs," *IEEE Design and Test of Computers*, Vol. 15, No. 2, pp. 14-21, April/June 1998.

[192] M. B. Kleiner, S. A. Kühn, P. Ramn, and W. Weber, "Thermal Analysis of Vertically Integrated Circuits," *Proceedings of the IEEE International Electron Devices Meeting*, pp. 487-490, December 1995.

[193] S. Im and K. Banerjee, "Full Chip Thermal Analysis of Planar (2-D) and Vertically Integrated (3-D) High Performance ICs," *Proceedings of the IEEE International Electron Devices Meeting*, pp. 727-730, December 2000.

[194] T.-Y. Chiang, S. J. Souri, C. O. Chui, and K. C. Saraswat, "Thermal Analysis of Heterogeneous 3-D ICs with Various Integration Scenarios," *Proceedings of the IEEE International Electron Devices Meeting*, pp. 681-684, December 2001.

[195] C. C. Liu, J. Zhang, A. K. Datta, and S. Tiwari, "Heating Effects of Clock Drivers in Bulk, SOI, and 3-D CMOS," *IEEE Transactions on Electron Device Letters*, Vol. 23, No. 12, pp. 716-728, December 2002.

[196] Z. Tan, M. Furmanczyk, M. Turowski, and A. Przekwas, "CFD-Micromesh: A Fast Geometrical Modeling and Mesh Generation Tool for 3D Microsystem Simulations," *Proceedings of the International Conference on Modeling and Simulation of Microsystems*, pp. 712-715, March 2000.

[197] P. Wilkerson, M. Furmanczyk, and M. Turowski, "Compact Thermal Model Analysis for 3-D Integrated Circuits," *Proceedings of the International*

Conference on Mixed Design of Integrated Circuits and Systems, pp. 277-282, June 2004.

[198] P. Wilkerson, M. Furmanczyk, and M. Turowski, "Fast, Automated Thermal Simulation of Three-Dimensional Integrated Circuits," *Proceedings of the Intersociety Conference on Thermal and Thermomechanical Phenomena in Electronic Systems*, pp. 706-713, June 2004.

[199] M. N. Sabry and H. Saleh, "Compact Thermal Models: A Global Approach," *Proceedings of the IEEE International Conference on Thermal Issues in Emerging Technologies*, pp. 33-39, January 2007.

[200] G. Digele, S. Lindenkreuz, and E. Kasper, "Fully Coupled Dynamic Electro-Thermal Simulation," *IEEE Transactions on Very Large Scale Integration (VLSI) Systems*, Vol. 5, No. 3, pp. 250-257, September 1997.

[201] S. Wunsche, C. Claub, and P. Schwarz, "Electro-Thermal Circuit Simulation Using Simulator Coupling," *IEEE Transactions on Very Large Scale Integration (VLSI) Systems*, Vol. 5, No. 3, pp. 277-282, September 1997.

[202] K. Puttaswamy and G. H. Loh, "Thermal Analysis of a 3-D Die Stacking High-Performance Microprocessor," *Proceedings of the ACM International Great Lakes Symposium on VLSI*, pp. 19-24, April/May 2006.

[203] J. Cong, J. Wei, and Y. Zhang, "A Thermal-Driven Floorplanning Algorithm for 3-D ICs," *Proceedings of the IEEE/ACM International Conference on Computer-Aided Design*, pp. 306-310, November 2004.

[204] W.-L. Hung *et al.*, "Interconnect and Thermal-Aware Floorplanning for 3-D Microprocessors," *Proceedings of the IEEE International Symposium on Quality Electronic Design*, pp. 98-103, March 2006.

[205] [Online]. Available: http://www.crhc.uiuc.edu/ACS/tools/ivm/about.html

[206] C. Addo-Quaye, "Thermal-Aware Mapping and Placement for 3-D NoC Designs," *Proceedings of the IEEE International SOC Conference*, pp. 25-28, September 2005.

[207] D. E. Goldberg, *Genetic Algorithms in Search, Optimization, and Machine Learning*, Addison-Wesley, 1989.

[208] H. Eisenmann and F. M. Johannnes, "Generic Global Placement and Floorplanning," *Proceedings of the IEEE/ACM Design Automation Conference*, pp. 269-274, June 1998.

[209] B. Goplen and S. Sapatnekar, "Efficient Thermal Placement of Standard Cells in 3-D ICs using a Force Directed Approach," *Proceedings of the IEEE/ACM International Conference on Computer-Aided Design*, pp. 86-89, November 2003.

[210] M. E. Van Valkenburg, *Network Analysis*, 3rd edition, Prentice-Hall, 1974.

[211] [Online]. Available: http://er.cs.ucla.edu/benchmarks/ibm-place

[212] [Online]. Available: http://www.cbl.ncsu.edu/pub/Benchmark_dirs/LayoutSynth92

[213] B. Goplen and S. Sapatnekar "Placement of Thermal Vias in 3-D ICs Using Various Thermal Objectives," *IEEE Transactions on Computer-Aided Design of Integrated Circuits and Systems*, Vol. 25, No. 4, pp. 692-709, April 2006.

[214] J. Cong and Y. Zhang, "Thermal Driven Multilevel Routing for 3-D ICs," *Proceedings of the IEEE Asia and South Pacific Design Automation Conference*, pp. 121-126, June 2005.

[215] J. Cong and Y. Zhang, "Thermal Via Planning for 3-D ICs," *Proceedings of the IEEE/ACM International Conference on Computer-Aided Design*, pp. 744-751, November 2005.

[216] J. Cong, M. Xie, and Y. Zhang, "An Enhanced Multilevel Routing System," *Proceedings of the IEEE/ACM International Conference on Computer-Aided Design*, pp. 51-58, November 2002.

[217] J. Cong, J. Fang, and Y. Zhang, "Multilevel Approach to Full-Chip Gridless Routing," *Proceedings of the IEEE/ACM International Conference on Computer-Aided Design*, pp. 234-241, November 2001.

[218] Z. Li *et al.*, "Efficient Thermal Via Planning Approach and its Application in 3-D Floorplanning," *IEEE Transactions on Computer-Aided Design of Integrated Circuits and Systems*, Vol. 26, No. 4, pp. 645-658, April 2007.

[219] T. H. Cormen, C. E. Leiserson, and R. L. Rivest, *Introduction to Algorithms*, The MIT Press, 1990.

[220] T. Zhang, Y. Zhang, and S. Sapatnekar, "Temperature-Aware Routing in 3-D ICs," *Proceedings of the IEEE Asia and South Pacific Design Automation Conference*, pp. 309-314, January 2006.

[221] J. A. Burns *et al.*, "A Wafer-Scale 3-D Circuit Integration Technology," *IEEE Transactions on Electron Devices*, Vol. 53, No. 10, pp. 2507-2515, October 2006.

[222] G. Chen and E. G. Friedman, "An *RLC* Interconnect Model Based on Fourier Analysis," *IEEE Transactions on Computer-Aided Design of Integrated Circuits and Systems*, Vol. 24, No. 2, pp. 170-183, February 2005.

[223] C. Ryu *et al.*, "High Frequency Electrical Circuit Model of Chip-to-Chip Vertical Via Interconnection for 3-D Chip Stacking Package," *Proceedings of the IEEE Topical Meeting on Electrical Performance of Electronic Packaging*, pp 151-154, October 2005.

[224] *Metal User's Guide, www.oea.com*

[225] P. Ramn *et al.*, "InterChip Via Technology for Vertical System Integration," *Proceedings of the IEEE International Interconnect Technology Conference*, pp. 160-162, June 2001.

[226] K. D. Boese *et al.*, "Fidelity and Near-Optimality of Elmore-Based Routing Constructions," *Proceedings of the IEEE International Conference on Computer Design*, pp. 81-84, October 1993.

[227] A. I. Abou-Seido, B. Nowak, and C. Chu, "Fitted Elmore Delay: A Simple and Accurate Interconnect Delay Model," *IEEE Transactions on Very Large Scale Integration (VLSI) Systems*, Vol. 12, No. 7, pp. 691-696, July 2004.

[228] J. P. Fishburn and C. A. Schevon, "Shaping a Distributed-*RC* Line to Minimize Elmore Delay," *IEEE Transactions on Circuits and Systems I: Fundamental Theory and Applications*, Vol. 42, No. 12, pp. 1020-1022, December 1995.

[229] J. Cong and K.-S. Leung, "Optimal Wiresizing under Elmore Delay Model," *IEEE Transactions on Computer-Aided Design of Integrated Circuits and Systems*, Vol. 14, No. 3, pp. 321-336, March 1995.

[230] J. D. Cho *et al.*, "Crosstalk-Minimum Layer Assignment," *Proceedings of the IEEE Conference on Custom Integrated Circuits*, pp. 29.7.1 - 29.7.4, May 1993.

[231] J. G. Ecker, "Geometric Programming: Methods, Computations and Applications," *SIAM Review*, Vol. 22, No. 3, pp. 338-362, July 1980.

[232] S. Boyd, S. J. Kim, L. Vandenberghe, and A. Hassibi, "A Tutorial on Geometric Programming," *Optimization and Engineering*, Vol. 8, No. 1, pp. 67-127, March 2007.

[233] Predictive Technology Model [Online]. Available: http://www.eas.asu.edu/~ptm

[234] W. Zhao and Y. Cao, "New Generation of Predictive Technology Model for Sub-45nm Design Exploration," *Proceedings of the IEEE International Symposium on Quality Electronic Design*, pp. 585-590, March 2006.

[235] J. Löfberg, "YALMIP: A Toolbox for Modeling and Optimization in MATLAB," *Proceedings of the IEEE International Symposium on Computer-Aided Control Systems Design*, pp. 284-289, September 2004.

[236] D. Henrion and J. B. Lasserre, "GloptiPoly: Global Optimization over Polynomials with Matlab and SeDuMi," *ACM Transactions on Mathematical Software*, Vol. 29, No. 2, pp. 165-194, June 2003.

[237] C.-P. Che, H. Zhou, and D. F. Wong, "Optimal Non-Uniform Wire-Sizing under the Elmore Delay Model," *Proceedings of the IEEE/ACM International Conference on Computer-Aided Design*, pp. 38-43, November 1996.

[238] A. M. Sule, *Design of Pipeline Fast Fourier Transform Processors Using 3 Dimensional Integrated Circuit Technology*, PhD Dissertation, North Carolina State University, December 2007.

[239] W. J. Dally, "Performance Analysis of k-ary n-cube Interconnection Networks," *IEEE Transaction on Computers*, Vol. 39, No. 6, pp. 775-785, June 1990.

[240] S. Palacharla, N. P. Jouppi, and J. E. Smith, "Complexity-Effective Superscalar Processors," *Proceedings of the IEEE International Conference on Computer Architecture*, pp. 206-218, June 1997.

[241] B. Vaidyanathan *et al.*, "Architecting Microprocessor Components in 3-D Design Space," *Proceedings of the IEEE International Conference on VLSI Design*, pp. 103-108, January 2007.

[242] R. P. Brent and H. T. Kung, "A Regular Layout for Parallel Adders," *IEEE Transactions on Computers*, Vol. C-31, No. 3, pp. 260-264, March 1982.

[243] P. M. Kogge and H. S. Stone, "A Parallel Algorithm for the Efficient Solution of a General Class of Recurrence Equations," *IEEE Transactions on Computers*, Vol. C-22, No. 8, pp. 786-793, August 1973.

[244] B. Black *et al.*, "Die Stacking (3D) Microarchitecture," *Proceedings of the IEEE/ACM International Symposium on Microarchitecture*, pp. 469-479, December 2006.

[245] S. S. Mukherjee *et al.*, "The Alpha 21364 Network Architecture," *IEEE Micro*, Vol. 22, No. 2, pp. 26-35, January/February 2002.

[246] Y. Xie, G. H. Loh, B. Black, and K. Bernstein, "Design Space Exploration for 3D Architectures," *ACM Journal on Emerging Technologies in Computing Systems*, Vol. 2, No. 2, pp. 65-103, April 2006.

[247] A. J. Smith, "Cache Memories," *ACM Computing Surveys*, Vol. 14, No. 3, pp. 473-530, September 1982.

[248] J. Sahuquillo and A. Pont, "Splitting the Data Cache: A Survey," *IEEE Concurrency*, Vol. 8, No. 3, pp. 30-35, July/September 2000.

[249] Y.-F. Tsai *et al.*, "Design Space Exploration for 3-D Cache," *IEEE Transactions on Very Large Scale Integration (VLSI) Systems*, Vol. 16, No. 4, pp. 444-455, April 2008.

[250] K. Zhang *et al.*, "A SRAM Design on 65 nm CMOS Technology with Integrated Leakage Reduction Scheme," *Proceedings of the IEEE International Symposium on VLSI Circuits*, pp. 294-295, June 2004.

[250] K. Zhang *et al.*, "A SRAM Design on 65 nm CMOS Technology with Integrated Leakage Reduction Scheme," *Proceedings of the IEEE International Symposium on VLSI Circuits*, pp. 294-295, June 2004.

[251] A. Zeng, J. Lü, K. Rose, and R. J. Gutmann, "First-Order Performance Prediction of Cache Memory with Wafer-Level 3D Integration," *IEEE Design and Test of Computers*, Vol. 22, No. 6, pp. 548-555, November/December 2005.

[252] S. J. E. Wilton and N. P. Jouppi, "CACTI: An Enhanced Cache Access and Cycle Time Model," *IEEE Journal of Solid-State Circuits*, Vol. 31, No. 5, pp. 677-688, May 1996.

[253] T. Sakurai, "Approximation of Wiring Delay in MOSFET LSI," *IEEE Journal of Solid-State Circuits*, Vol. SC-18, No. 4, pp. 418-426, August 1983.

[254] M. Mamidipaka, K. Khouri, N. Dutt, and M. Abadir, "Analytical Models for Leakage Power Estimation of Memory Array Structures," *Proceedings of the IEEE/ACM International Conference on Hardware/Software Codesign and System Synthesis*, pp. 146-151, September 2004.

[255] G. M. Link and N. Vijaykrishnan, "Thermal Trends in Emergent Technologies," *Proceedings of the IEEE International Symposium on Quality Electronic Design*, pp. 625-632, March 2006.

[256] M. B. Kleiner, S. A. Kühn, P. Ramn, and W. Weber, "Performance Improvement of the Memory Hierarchy of RISC-Systems by Application of 3-D Technology," *IEEE Transactions on Components, Packaging, and Manufacturing Technology – Part B*, Vol. 19, No. 4, pp. 709-718, November 1996.

[257] D. H. Albonesi and I. Koren, "Improving the Memory Bandwidth of Highly-Integrated, Wide-Issue, Microprocessor-Based Systems," *Proceedings of the IEEE International Conference on Paraller Architectures and Compilation Techniques*, pp. 126-135, November 1997.

[258] K. Suzuki *et al.*, "A 500 MHz, 32 bit, 0.4 μm CMOS RISC Processor," *Journal of Solid-State Circuits*, Vol. 29, No. 12, pp. 1464-1473, December 1994.

[259] C. E. Gimarc and V. M. Milutinovic, "A Survey of RISC Prcocessors and Computers of the Mid-1980s," *Computer*, Vol. 20, No. 9, pp. 59-69, September 1987.

[260] Intel, [Online]. Available: http://www.intel.com/products/processor/core2/index.htm

[261] G. H. Loh, Y. Xie, and B. Black, "Processor Design in 3D Die-Stacking Technologies," *IEEE Micro*, Vol. 27, No. 3, pp. 31-48, May/June 2007.

[262] Z. Guz, I. Keidar, A. Kolodny, and U. C. Weiser, "Nahalal: Cache Organization for Chip Multiprocessors," *Computer Architecture Letters*, Vol. 6, No. 1, pp. 21-24, January 2007.

[263] D. Bertozzi *et al.*, "NoC Synthesis Flow for Customized Domain Specific Multiprocessor Systems-on-Chip," *IEEE Transactions on Parallel and Distributed Systems*, Vol. 16, No. 2, pp. 113-129, February 2005.

[264] J. C. Koob *et al.*, "Design of a 3-D Fully Depleted SOI Computational RAM," *IEEE Transactions on Very Large Scale Integration (VLSI) Systems*, Vol. 13, No. 3, pp. 358-368, March 2005.

[265] S. Kumar *et al.*, "A Network on Chip Architecture and Design Methodology," *Proceedings of the IEEE International Annual Symposium on VLSI*, pp. 105-112, April 2002.

[266] V. F. Pavlidis and E. G. Friedman, "3-D Topologies for Networks-on-Chip," *Proceedings of the IEEE International SOC Conference*, pp. 285-288, September 2006.

[267] V. F. Pavlidis and E. G. Friedman, "3-D Topologies for Networks-on-Chip," *IEEE Transactions on Very Large Scale Integration (VLSI) Systems*, Vol. 15, No. 10, pp. 1081-1090, October 2007.

[268] F. Li *et al.*, "Design and Management of 3D Chip Multiprocessors Using Network-in-Memory," *Proceedings of the IEEE International Symposium on Computer Architecture*, pp. 130-142, June 2006.

[269] A. Jantsch and H. Tenhunen, *Networks on Chip*, Kluwer Academic Publishers, 2003.

[270] M. Millberg *et al.*, "The Nostrum Backbone - A Communication Protocol Stack for Networks on Chip," *Proceedings of the IEEE International Conference on VLSI Design*, pp. 693-696, January 2004.

[271] J. M. Duato, S. Yalamanchili, and L. Ni, *Interconnection Networks: An Engineering Approach*, Morgan Kaufmann, 2003.

[272] W. J. Dally and B. Towles, *Principles and Practices of Interconnection Networks*, Morgan Kaufmann, 2004.

[273] L.-S. Peh and W. J. Dally, "A Delay Model for Router Microarchitectures," *IEEE Micro*, Vol. 21, No. 1, pp. 26-34, January/February 2001.

[274] T. Sakurai, "Closed-Form Expressions for Interconnection Delay, Coupling, and Crosstalk in VLSIs," *IEEE Transactions on Electron Devices*, Vol. 40, No. 1, pp. 118-124, January 1993.

[275] T. Sakurai and A. R. Newton, "Alpha-Power Law MOSFET Model and its Applications to CMOS Inverter Delay and other Formulas," *IEEE Journal of Solid-State Circuits*, Vol. 25, No. 2, pp. 584-594, April 1990.

[276] G. Chen and E. G. Friedman, "Low-Power Repeaters Driving RC and RLC Interconnects With Delay and Bandwidth Constraints," *IEEE Transactions on Very Large Integration (VLSI) Systems*, Vol. 12, No. 2, pp. 161-172, February 2006.

[277] Y. I. Ismail, E. G. Friedman, and J. L. Neves, "Equivalent Elmore Delay for RLC trees," *IEEE Transactions on Computer-Aided Design of Integrated Circuits and Systems*, Vol. 19, No. 1, pp. 83-97, January 2000.

[278] Y. I. Ismail, E. G. Friedman, and J. L. Neves, "Figures of Merit to Characterize the Importance of On-Chip Inductance," *IEEE Transactions on Very Large Scale Integration (VLSI) Systems*, Vol. 7, No. 4, pp. 442-449, December 1999.

[279] K. Banerjee and A. Mehrotra, "A Power-Optimal Repeater Insertion Methodology for Global Interconnects in Nanometer Design," *IEEE Transactions on Electron Devices*, Vol. 49, No. 11, pp. 2001-2007, November 2002.

[280] H. J. M. Veendrick, "Short-Circuit Dissipation of Static CMOS Circuitry and its Impact on the Design of Buffer Circuits," *IEEE Journal of Solid-State Circuits*, Vol. SC-19, No. 4, pp. 468-473, August 1984.

[281] K. Nose and T. Sakurai, "Analysis and Future Trend of Short-Circuit Power," *IEEE Transactions on Computer-Aided Design of Integrated Circuits and Systems*, Vol. 19, No. 9, pp. 1023-1030, September 2000.

[282] G. Chen and E. G. Friedman, "Effective Capacitance of RLC Loads for Estimating Short-Circuit Power," *Proceedings of the IEEE International Symposium on Circuits and Systems*, pp. 2065-2068, May 2006.

[283] P. R. O'Brien and T. L. Savarino, "Modeling the Driving-Point Characteristic of Resistive Interconnect for Accurate Delay Estimation," *Proceedings of the IEEE/ACM International Conference on Computer-Aided Design*, pp. 512-515, April 1989.

[284] H. Wang, L.-S. Peh, and S. Malik, "Power-Driven Design of Router Microarchitectures in On-Chip Networks," *Proceedings of the IEEE International Symposium on Microarchitecture*, pp. 105-116, December 2003.

[285] C. Marcon *et al.*, "Exploring NoC Mapping Strategies: an Energy and Timing Aware Technique," *Proceedings of the IEEE/ACM Design, Automation and Test in Europe Conference and Exhibition*, Vol. 1, pp. 502-507, March 2005.

[286] P. P. Pande *et al.*, "Performance Evaluation and Design Trade-Offs for Network-on-Chip Interconnect Architectures," *IEEE Transactions on Computers*, Vol. 54, No. 8, pp. 1025-1039, August 2005.

[287] R. Marculescu, U. Y. Ogras, and N. H. Zamora, "Computation and Communication Refinement for Multiprocessor SoC Design: A System-Level Perspective," *ACM Transactions on Design Automation of Electronic Systems*, Vol. 11, No. 3, pp. 564-592, July 2006.

[288] V. Soteriou, H. Wang, and L.-S. Peh, "A Statistical Trace Model for On-Chip Interconnection Networks," *Proceedings of the IEEE International Symposium on Modeling, Analysis, and Simulation of Computer and Telecommunication Systems*, pp. 104-116, September 2006.

[289] K. Siozios, K. Sotiriadis, V. F. Pavlidis, and D. Soudris, "Exploring Alternative 3D FPGA Architectures: Design Methodology and CAD Tool Support," *Proceedings of the IEEE International Conference on Field Programmable Logic and Applications*, pp. 652-655, August 2007.

[290] G.-M. Chiu, "The Odd-Even Turn Model for Adaptive Routing," *IEEE Transactions on Parallel and Distributed Systems*, Vol. 11, No. 7, pp. 729-738, July 2000.

[291] K. Lahiri *et al.*, "Evaluation of the Traffic-Performance Characteristics of System-on-Chip Communication Architectures," *Proceedings of the Conference on VLSI Design*, pp. 29-35, October 2000.

[292] T. T. Ye, L. Benini, and G. De Micheli, "Analysis of Power Consumption on Switch Fabrics in Network Routers," *Proceedings of the IEEE/ACM Design Automation Conference*, pp. 524-529, June 2002.

[293] B. Feero and P. P. Pande, "Performance Evaluation for Three-Dimensional Networks-on-Chip," *Proceedings of the IEEE International Symposium on VLSI*, pp. 305-310, March 2007.

[294] [Online]. Available: http://www.xilinx.com

[295] M. J. Alexander *et al.*, "Placement and Routing for Three-Dimensional FPGAs," *Proceedings of the Canadian Workshop on Field-Programmable Devices*, pp. 11-18, May 1996.

[296] [Online]. Available: http://www.xilinx.com/products/ silicon_solutions/fpgas/ spartan_series /spartan3_ fpgas/index.htm

[297] A. Rahman, S. Das, A. P. Chandrakasan, and R. Reif, "Wiring Requirement and Three-Dimensional Integration Technology for Field Programmable Gate Arrays," *IEEE Transactions on Very Large Scale Integration (VLSI) Systems*, Vol. 11, No. 1, pp. 44-53, February 2003.

[298] G. G. Lemieux and S. D. Brown, "A Detailed Routing Algorithm for Allocating Wire Segments in Field-Programmable Gate Arrays," *Proceedings of the IEEE Physical Design Workshop*, pp. 215-226, April 1993.

[299] V. Betz and J. Rose, "VPR: A New Packing, Placement, and Routing Tool for FPGA Research," *Proceedings of the International Workshop on Field Programmable Logic Applications*, pp. 213-222, September 1997.

[300] M. J. Alexander, J. P. Cohoon, J. L. Ganley, and G. Robins, "Performance-Oriented Placement and Routing for Field-Programmable Gate Arrays," *Proceedings of the IEEE International European Design Automation Conference*, pp. 80-85, September 1995.

[301] M. J. Alexander *et al.*, "Three-Dimensional Field-Programmable Gate Arrays," *Proceedings of the IEEE International ASIC Conference*, pp. 253-256, September 1995.

[302] M. J. Alexander, J. P. Cohoon, J. L. Ganley, and G. Robins, "An Architecture-Independent Approach to FPGA Routing Based on Multi-Weighted Graphs," *Proceedings of the IEEE International European Design Automation Conference*, pp. 259-264, September 1995.

[303] C. Ababei, H. Mogal, and K. Bazargan, "Three-Dimensional Place and Route for FPGAs," *Proceedings of the IEEE Asia and South Pacific Design Automation Conference*, pp. 773-778, January 2005.

[304] V. Betz, J. Rose, and A. Marquardt, *Architecture and CAD for Deep-Submicron FPGAs*, Kluwer Academic Publishers, 1999.

[305] G. Karypis, R. Aggarwal, V. Kumar, and S. Shekhar, "Multilevel Hypergraph Partitioning: Applications in VLSI Domain," *IEEE Transactions on Very Large Scale Integration (VLSI) Systems*, Vol. 7, No.1, pp. 69-79, March 1999.

[306] P. Maidee, C. Ababei, and K. Bazargan, "Fast Timing-Driven Partitioning Based Placement for Island Style FPGAs," *Proceedings of the IEEE/ACM Design Automation Conference*, pp. 598-603, June 2003.

[307] C. Ebeling, L. McMurchie, S. A. Hauck, and S. Burns, "Placement and Routing Tools for the Triptych FPGAs," *IEEE Transactions on Very Large Scale Integration (VLSI) Systems*, Vol. 3, No. 4, pp. 472-483, December 1995.

[308] K. Siozios, K. Sotiriadis, V. F. Pavlidis, and D. Soudris, "A Software-Supported Methodology for Designing High-Performance 3D FPGA Architectures," *Proceedings of the IFIP International Conference on Very Large Scale Integration*, pp. 54-59, October 2007.

[309] K. Siozios *et al.*, "A Novel FPGA Architecture and an Integrated Framework of CAD Tools for Implementing Applications," *IEICE Transactions on Information and Systems*, Vol. E88-D, No. 7, pp. 1369-1380, July 2005.

[310] C. Ababei *et al.*, "Placement and Routing in 3D Integrated Circuits," *IEEE Design and Test of Computers*, Vol. 22, No. 6, pp. 520-531, November/December 2005.

[311] M. Lin, A. El Gamal, Y.-C. Lu, and S. Wong, "Performance Benefits of Monolithically Stacked 3-D FPGA," *IEEE Transactions on Computer-Aided Design of Integrated Circuits and Systems*, Vol. 26, No. 2, pp. 216-229, February 2007.

[312] E. G. Friedman (Ed.), *Clock Distribution Networks in VLSI Circuits and Systems*, New Jersey, IEEE Press, 1995.

[313] D. W. Bailey and B. J. Benschneider, "Clocking Design and Analysis for a 600-MHz Alpha Microprocessor," *IEEE Journal of Solid-State Circuits*, Vol. 22, No. 11, pp. 1627-1633, November 1998.

[314] T. Xanthopoulos *et al.*, "The Design and Analysis of the Clock Distribution Network for a 1.2 GHz Alpha Microprocessor," *Proceedings of the IEEE International Solid-State Circuits Conference*, pp. 402-402, February 2001.

[315] H. B. Bakoglu, *Circuits, Interconnections, and Packaging for VLSI*, Addison-Wesley, 1990.

[316] V. Pavlidis and E. G. Friedman, "Interconnect Delay Minimization through Interlayer Via Placement in 3-D ICs," *Proceedings of the ACM Great Lakes Symposium on VLSI*, pp. 20-25, April 2005.

[317] W. Cui, H. Chen, and Y. Han, "VLSI Implementation of Universal Random Number Generator," *Proceedings of the IEEE Asia-Pacific Conference on Circuits and Systems*, Vol. 1, pp. 465-470, October 2002.

[318] A. Deutsch and P. J. Restle, "Designing the Best Clock Distribution Network," *Proceedings of the IEEE Symposium on VLSI Circuits*, pp. 2-5, June 1998.

[319] E. G. Friedman, "Clock Distribution Networks in Synchronous Digital Integrated Circuits," *Proceedings of the IEEE*, Vol. 89, No. 5, pp. 665-692, May 2001.

[320] N. Hedenstierna and K. O. Jeppson, "CMOS Circuit Speed and Buffer Optimization," *IEEE Transactions on Computer-Aided Design*, Vol. CAD-6, No. 2, pp. 270-281, March 1987.

[321] N. C. Li, G. L. Haviland, and A. A. Tuszynski, "CMOS Tapered Buffer," *IEEE Journal of Solid-State Circuits*, Vol. 25, No. 4, pp. 1005-1008, August 1990.

[322] C. Punty and L. Gal, "Optimum Tapered Buffer," *IEEE Journal of Solid-State Circuits*, Vol. 27, No. 1, pp. 1005-1008, January 1992.

[323] B. S. Cherkauer and E. G. Friedman, "A Unified Design Methodology for CMOS Tapered Buffers," *IEEE Transactions on Very Large Scale Integration (VLSI) Systems*, Vol. 3, No. 1, pp. 99-111, March 1995.

Index

Printed and bound by CPI Group (UK) Ltd, Croydon, CR0 4YY

03/10/2024

01040314-0010